VII International Scientific and Practical Conference on Innovations in Mechanical Engineering (ISPCIME-2015)

IOP Conference Series: Materials Science and Engineering
Volume 126

Kemerovo, Russia
23 – 25 September 2015

Editor:

Valeriy Blumenstein

ISBN: 978-1-5108-2325-9
ISSN: 1757-8981

Printed from e-media with permission by:

Curran Associates, Inc.
57 Morehouse Lane
Red Hook, NY 12571

Some format issues inherent in the e-media version may also appear in this print version.

Printed by Curran Associates, Inc. (2016)

For permission requests, please contact the Institute of Physics
at the address below.

Institute of Physics
Dirac House, Temple Back
Bristol BS1 6BE UK

Phone: 44 1 17 929 7481
Fax: 44 1 17 920 0979

techtracking@iop.org

Additional copies of this publication are available from:

Curran Associates, Inc.
57 Morehouse Lane
Red Hook, NY 12571 USA
Phone: 845-758-0400
Fax: 845-758-2633
Email: curran@proceedings.com
Web: www.proceedings.com

Table of contents

Volume 126

VII International Scientific and Practical Conference on Innovations in Mechanical Engineering (ISPCIME-2015)
23–25 September 2015, Kemerovo, Russia

Accepted papers received: 5 April 2016
Published online: 27 April 2016

Preface

Papers

IOP Conf. Series: Materials Science and Engineering **126** (2016) 011001 doi:10.1088/1757-899X/126/1/011001

Preface

The Conference is devoted to the discussion of the actual problems of the mechanical engineering industry, the strategic branch without which a stable, steady, and dynamically developing economy is impossible. The purpose of the conference is granting a meeting place for the scientists and specialists of the leading Russian and foreign higher education, scientific and research institutes, enterprises where they can present their scientific and practical achievements, debate about problems of innovations, exchange experience in the training of specialists, discuss status and ways of improving the plan system as well as management and technological processes in mechanical engineering.

Valeriy Blumenstein
Doctor of Technical Sciences, Professor
T.F. Gorbachev Kuzbass State Technical University,
Russian Federation

IOP Conf. Series: Materials Science and Engineering **126** (2016) 011002 doi:10.1088/1757-899X/126/1/011002

Peer review statement

All papers published in this volume of *IOP Conference Series: Materials Science and Engineering* have been peer reviewed through processes administered by the proceedings Editors. Reviews were conducted by expert referees to the professional and scientific standards expected of a proceedings journal published by IOP Publishing.

Analysis of the Face Milling Process Based on the Imitation Modelling

V A Khomenko [1,a], A O Cherdancev [1,b], P O Cherdancev [1,c], V D Goncharov [1,d], A Kulawik [2,e]

[1] Polzunov Altai State Technical University, Lenina 46, 656038, Barnaul, Russian Federation
[2] Czestochowa University of Technology, Armii Krajowej 2142-201 Czestochowa, Poland

[a] agtu-tap@mail.ru, [b] hypertigr@mail.ru, [c] p004092@yandex.ru, [d] agtu-tap@mail.ru, [e] adam.kulawik@icis.pcz.pl

Abstract. The results of simulation modeling of the face milling process allowed to analyze the surface finish dependence on cutting conditions. It is noted that the surface irregularity in the form of unevenness appears in face milling. Besides, the ratio of feed rate to spindle revolutions speed influences dimensions of surface roughness. The data received are experimentally confirmed.

1. Introduction

Engineering support of the parts of required quality along with a simultaneous costs reduction of machining processes is one of the most important problems of mechanical engineering. Geometrics of milled surface quality, in particular, surface roughness, influence almost all part performances. To measure surface roughness, the parameters of height, step and shape are used. If all the aspects of the industrial process system are taken into consideration, cutting conditions and structural-geometric parameters of cutting tools are the main factors in making topography of the milled surface.

Simulation modeling is the most effective research method in such complex process as shaping by face milling. Simulated model allows to characterize the development of a prototype process in time, and this makes possible to carry out computational experiments. Computational experiments allow analyzing surface roughness dependence on different milling process parameters; moreover, unlike real experiments there is no demand in a large amount of financial and time spending [1-9].

Consequently, a conceptual foresight of the surface finish by using simulation modeling in face milling is actual. Design technique of face milling operations can be based on it.

2. Materials and research methods

For carrying out numerical experiments we have created a simulated model of the face milling process. The model structure is shown in Figure 1.

The model consists of the following models: the model of the milled surface; the models of the cutting edges; the model of the mill; the model of the milling kinematics; the model of the cutting process; the model of the plastic deformations; the model of the cutting forces; the model of the vibrations; the model of the teeth wear. All these models are connected with each other by information streams and interrelate at every step of iterative computation in time.

IOP Conf. Series: Materials Science and Engineering **126** (2016) 012001 doi:10.1088/1757-899X/126/1/012001

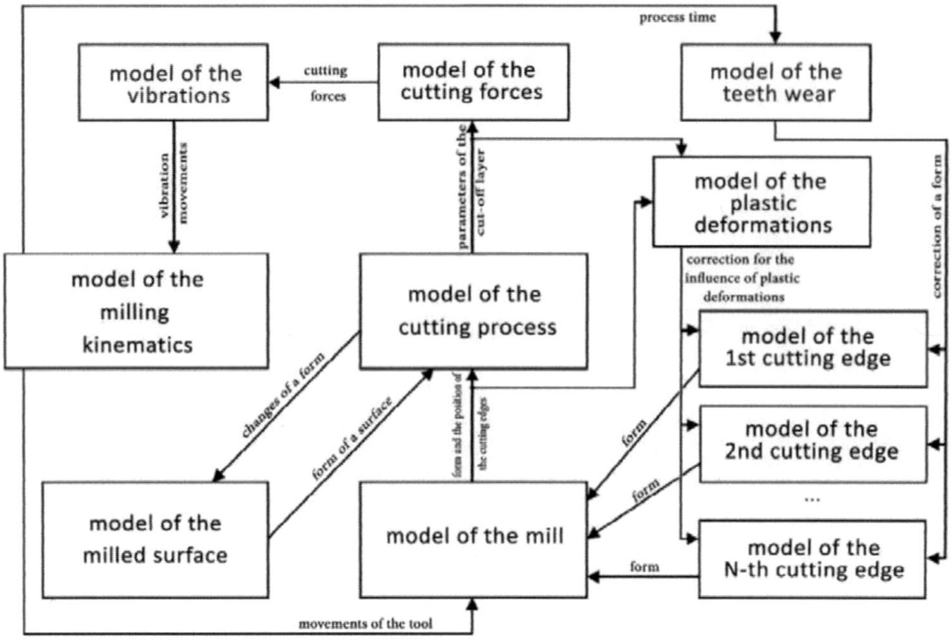

Fig. 1 - Simulated model structure

The modeling result can be illustrated as a topographic scheme of micro irregularities. This scheme demonstrates all the information about geometric parameters of the surface finish.

On the basis of the model, numerical experiments were carried out aimed at studying the connection between geometric parameters of the surface finish and structural-geometric parameters of cutting tools and cutting conditions in face milling.

The results have shown that the general principles of formation of the micro relief on the milled surface stay the same for various cutting conditions, milling cutters constructions and machined materials. The following are the results received by modeling the milling process of steel 1045 (AISI) with a face milling cutter having the diameter 50 mm, 4 teeth, SCEA=23°, γr=5°, γa=5°, ε=108°, r=0.8 mm, the cutting head material is T15C6 (a titanium-tungsten alloy), feed per minute equals 1600 mm/min, spindle speed is 400 revolutions per minute, cutting depth is 1 mm. The cutting conditions are chosen for efficient analyzing of the principles and demonstrating research results.

In topogram of the machined surface (Figure 2) ordinate axis is in line with feed directions, the range on the abscise axis illustrates milling width. The values of micro irregularities are coloured.

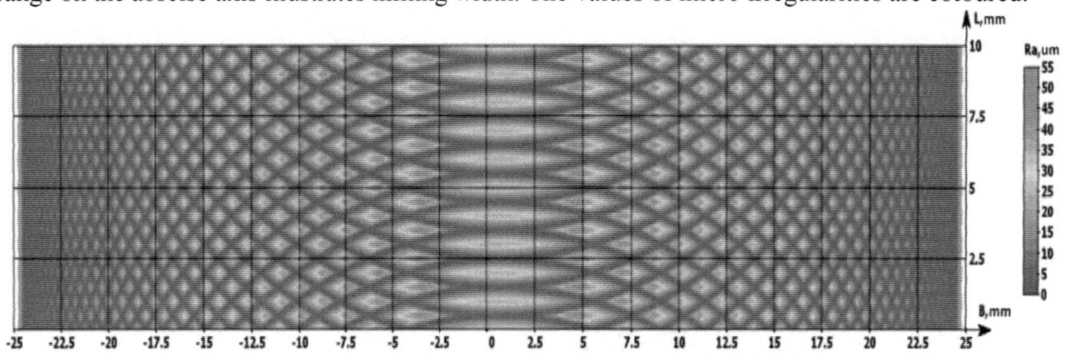

Fig. 2 – Topogram of the machined surface

Analysis of this topogram lets us conclude that different values will be obtained for different positions of roughness measure lines. This is illustrated in Figire 3, which demonstrates the relation between height parameters (ordinate axis) measured along the lines of parallel feeds and the position of the cross-section towards the full milling width (abscise axis).

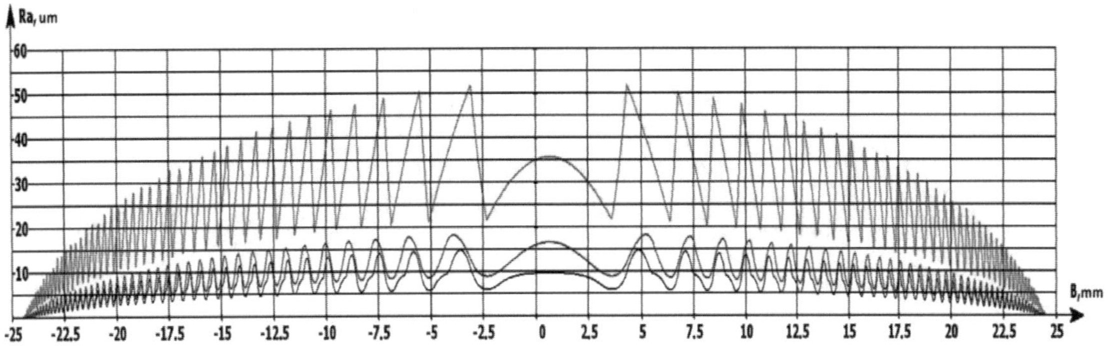

Fig. 3 – Height parameters of surface roughness (vertically down: Rmax, m, Ra)

Analysis of the figure shows that the surface receives a form deviation which is determined by the character of face milling. This deviation is an unevenness which can be calculated as the difference between the ordinate of the cross-cup the envelop curve and cutting width border line and the maximum value of the envelop curve. Another characteristic of the topography produced by face milling process is improving the surface finish in the peripheral area; it can be explained with the influence of the cutting kinematic and better conditions while milling if the layer cut off peripherally is thinner.

One more important parameter of surface finish is a relative carrying length of the profile at some level. The scheme of setting up the values of the relative carrying length of the profile in dependence to the cross-sectional position and level p (Figure 4) illustrates significant changes of this parameter for relatively close set up cross-sections.

Arithmetic average roughness height depends on both the position of profile line and its direction (Figure 5). The maximum values of the arithmetic deviation have the edge positions of the central area (if measured parallel to feed motion).

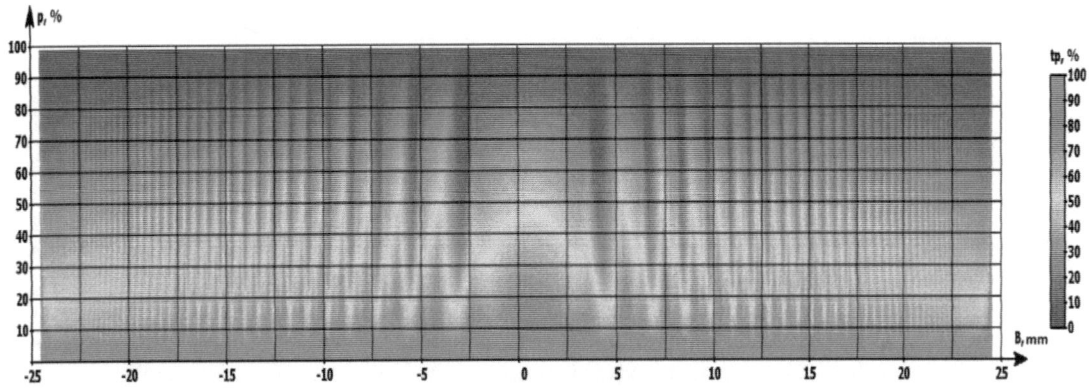

Fig. 4 – Setting up the relative carrying length of the profile

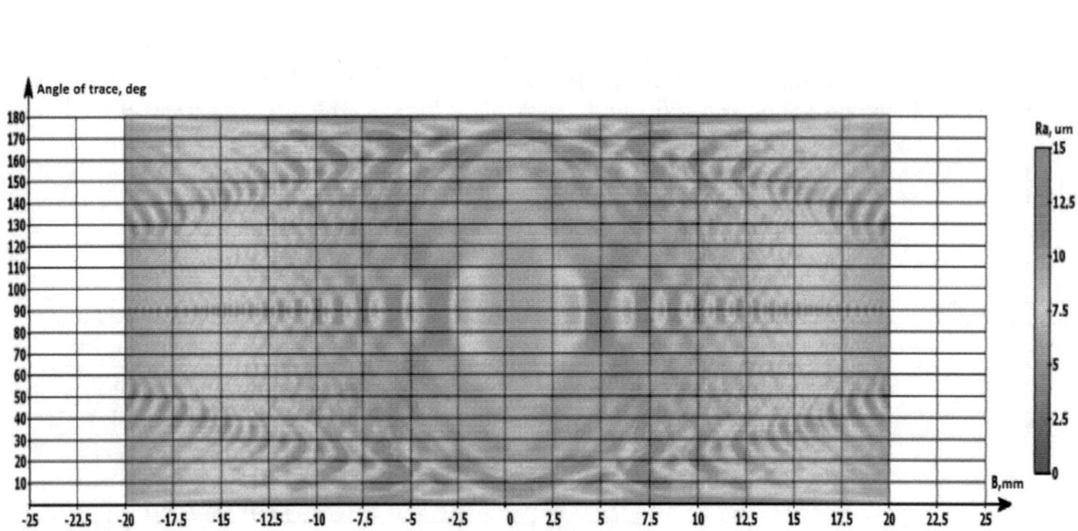

Fig. 5- Setting up the arithmetic average roughness height

The central area of the surface produced by face milling is likely to have rather different roughness with a relatively slightly varying value of feed to spindle revolutions per minute ratio (Figure 6).

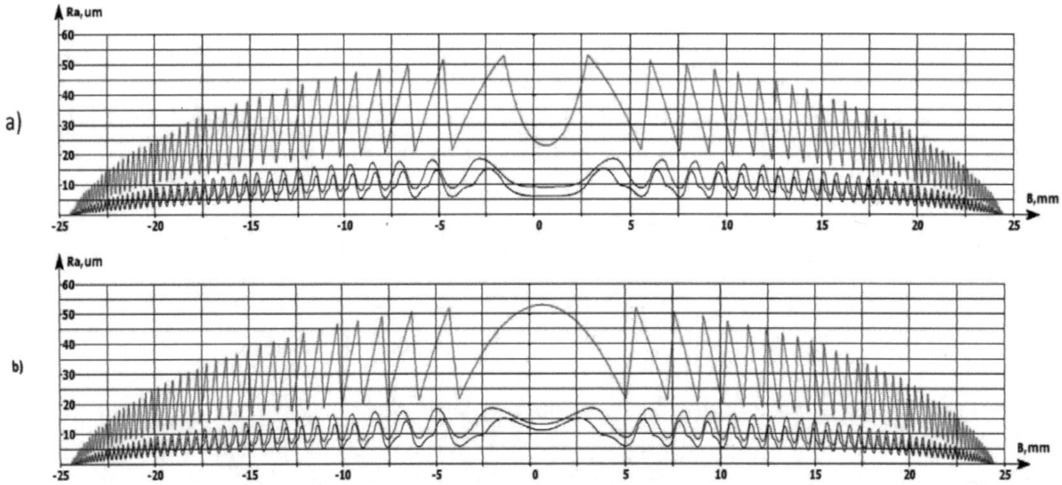

Fig. 6 – Height parameters with n=400 rpm for different feed values: a) S=1612 mm/min; b) S=1619 mm/min

Dependency graph of arithmetic average roughness height from feed per minute for cross-sectional position with B=0 mm shows that periodic coincidences of teeth motion trajectory in case of "backward" motion and previously made machining signs influence the character of the obtained dependence, it is expressed by local maximum height parameters (Figure 7). The local minimums correspond with a teeth pass in case of "backward" motion through the rest of micro irregularities.

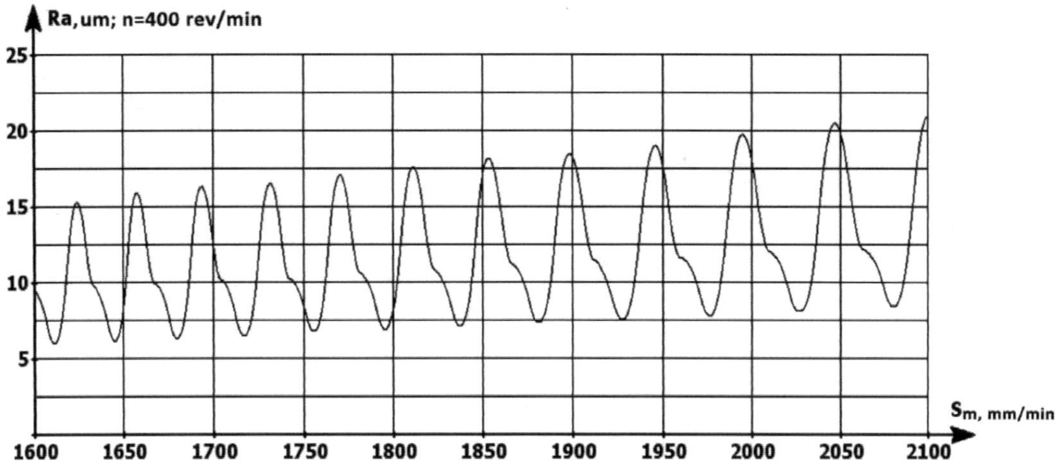

Fig. 7 – Dependency Ra in the central area from feed per minute value

Dependency of the maximum value Ra from feed per tooth which is shown in Figure 8, attracts a technologist in practical terms most of all.

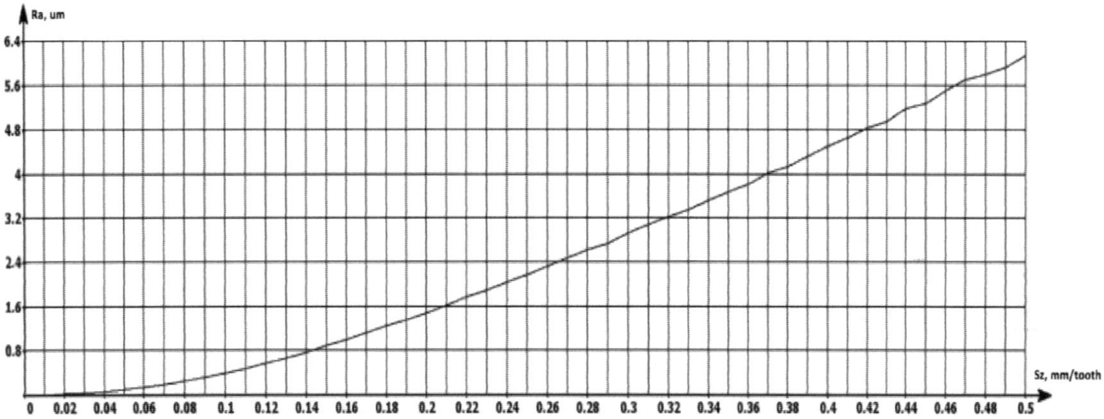

Fig. 8 – Dependency of the maximum value Ra from feed per tooth

This dependency is received due to mathematical analysis of the results of numerous numerical experiments conducted with the use of the developed model.

3. Experimental technique

To prove the possibility of practical using the results received while the work with the simulation model, a series of experiments was conducted under different cutting conditions with the cutters of different construction. The experiment was conducted on the milling machine GF2171 equipped with CNC FMS 3200, 24 pieces made of steel 1045 (AISI) were milled with cutting depth 1 mm. Roughness measurements were made with roughness measuring station Hommel Tester W55.

Table 1 – Experiment conditions and results

Cutting tool	Feedrate (mm/min)	RPM (min^{-1})	Ra (um)				tp (%)			
			Measured	Average	Calculated	Deviation (%)	Measured	Average	Calculated	Deviation (%)
Mill Ø50, 4 teeth, SCEA=23°, T15C6	F=630	S=800	1.39				33.8			
			1.47	1.44	1.46	1.4	32.9	33.4	31.4	5.9
			1.45				33.4			
		S=1600	0.45				29.9			
			0.43	0.42	0.38	9.5	30.5	30.3	29.0	4.3
			0.39				30.8			
	F=1000	S=800	2.97				36.2			
			3.03	3.01	3.09	2.7	36.9	36.1	33.6	6.9
			3.01				35.2			
		S=1600	0.90				30.7			
			0.89	0.90	0.96	6.7	30.9	31.0	29.6	4.5
			0.92				31.4			
Mill Ø100, 8 teeth, SCEA=23°, T15C6	F=630	S=400	1.36				31.4			
			1.42	1.39	1.45	4.3	31.9	31.2	30.4	2.5
			1.40				30.2			
		S=800	0.39				29.6			
			0.33	0.36	0.38	5.5	29.9	30.1	28.9	3.9
			0.37				30.9			
	F=1000	S=400	3.24				37.3			
			3.18	3.22	3.10	3.7	36.4	36.8	35.2	4.3
			3.25				36.8			
		S=800	0.94				29.8			
			0.99	0.95	0.96	1.1	31.2	30.6	29.6	3.3
			0.92				30.8			

The data received were mathematically analyzed and as a result the values of the arithmetic average roughness height and the relative carrying length of the profile were obtained. The data analysis showed that the differences of experimental and calculated by the model parameters are less than or equal 10 %, and the average value of deviations makes up about 5 %. Thus, the developed model fully affords the imitation of milling cutters operation.

4. Conclusion

1. Topogram analysis reveals that different positions of profile lines depend on far different values of roughness parameters.
2. Mathematical model provides the calculation of the surface roughness parameters rather accurate. The model has revealed the principles which illustrate as sufficient as possible the process of the surface topography formation in face milling according to a strict technological system.
3. Geometrical parameters dependencies of a part surface finish from milling input parameters received in the result of the numerical experiments can be used in design of milling operations by machining various parts.

Reference

[1] V.A. Khomenko, A.O. Cherdantsev, Stochastic simulation modelling of forming geometrical characteristics of surface in face milling. Polzunovskij vestnik, Barnaul, 2012, 324-328.
[2] V.A. Khomenko, A.O. Cherdantsev, Modelling of formation of Geometric parameters of face milled surface quality, 2nd International scientific and practical conference «Innovations in mechanical engineering» (ISPCIME-2011), 237-240.

[3] V.A. Khomenko, A.O. Cherdantsev, P.O. Cherdantsev, Optimizaton of face milling operation by maximum perfomance criterion based on imitation simulation. Polzunovskij vestnik, Barnaul, 2015, 49-54.
[4] A.M. Markov, Improving performance of face milling operations on CNC machines by controlling deviations of form and waviness of flat body parts. Dis. Ph.D. Chelyabinsk, Chelyabinsk State University, 1993.
[5] V.M. Svinin, Oscillation simulation of technological system for face milling with modulated cutting speed. High technologies in mechanical engineering., materials of all-Russian Scientific and Technical Conference - Samara, 2007, 187-189.
[6] High-performance metal cutting. Moscow, Publishing House "Polygraphy", 2003.
[7] V.V. Kuvshinskii, Milling. Moscow, Publishing House "Machinary", 1977.
[8] S.L. Leonov, E.Yu. Tatarkin, Construction of models for the calculation of roughness and topography of the machined surfaces. Polzunovskij vestnik, Barnaul, 2008, 170-174.
[9] A.G. Suslov, Technological parameters providing of the parts' surface layer. Publishing House "Machinary", 1978.

Quality Improvement of Chrome-Diamond Coatings on Flowing Chrome Plating

V N Belyaev, A Yu Koslyuk, A V Lobunets, A S Andreyev

JSC Federal Research & Production Center ALTAI, Russia, Biysk, Socialisticheskaya, 1

E-mail: lab.nanodiamond@rambler.ru

Abstract. The research results of the process of flowing chrome plating of internal surfaces of long-length cylindrical articles with the usage of electrolyte with ultra-dispersed diamonds when continuous article rotation, while chromium-plating, are presented. During experiments the following varying technological parameters: electrolyte temperature and article frequency rotation were chosen, and experimental samples were obtained. Estimation of porosity, micro-hardness, thickness of chrome coatings and uniformity were performed as well as the precipitation structure by the method of scanning electron microscopy. The results showed that the use of ultra-dispersed diamonds and realization of the scheme with rotation of detail-cathode when flowing chromium-plating allows one to increase servicing characteristics of the coating due to the decrease of grains size of chrome coating and porosity, and due to the increase of micro-hardness, so confirming the efficiency of using the suggested scheme of coating application and the given type of ultra-dispersed fillers when chromium-plating.

Introduction

High hardness, low friction coefficient, heat-resistance and high chemical stability make chrome-electroplated parts high wear-resistant. Chrome plating is widely used for improving wear-resistance of different articles, including cylindrical parts of inner friction surface with large length/diameter ratio. The process of electroplating in flow-through electrolyte, or flowing chrome-plating [1], is used for these long-sized articles.

Special set-ups providing forced supply of electrolyte in space between surfaces of the part to be electroplated and anode are used for chrome plating in flowing electrolyte. Forced supply of electrolyte provides its continuous substitution and uniform gas saturation in inter-electrode volume. It is advisable that these setups should be applied for chrome plating of inner surfaces of long articles when electrolyte saturation with the formed gases is too high that it prevents the standard process from depositing chrome. Maximum current efficiency (about 20%) is achieved on flowing chrome plating at a flowing electrolyte speed of 1 m/sec. The deposit is smoothed in flowing electrolyte making it possible to obtain solid glittering coatings or milk-white wear-resistant coatings at high current densities.

The most important problem both on chrome electroplating and depositing other coatings is achievement of uniform parameters, including geometric ones, along the surface and the volume of coating. It is especially actual for the chrome plating process as the change in deposition conditions along the inner surface of an article is added to the worst among other electrolytes dispersion ability.

It is necessary that stable electrolyte flow with the predetermined speed of near-surface anode and cathode layers should be firstly generated to get uniform quality parameters of chrome plating of inner surfaces along the length of long articles. At the same time, it is necessary to provide process parameters of chrome plating, airtightness of the set-up connection units with an article and the ultimate fast and complete removal of hydrogen from the deposition area [2-4]. The use of electrolytes with ultra-dispersed diamonds (UDD) and continuous article rotation during chrome plating instead of low-concentrated electrolytes recommended for flow-through chrome plating, make it possible to get high-quality chrome-electrochemical coatings. It gives higher micro-hardness parameters, less irregularity of geometrical parameters of chrome deposits, low number of pores per unit of surface area and what is more to get the necessary intensity of the process [5,6] due to increased current densities.

Experimental procedure (part)

The diagram of flowing chrome plating is given in Figure 1. The flowing chrome plating set-up using UDD consists of an article 1 to be rotated during chrome plating relative to its axis, header 2 and drain units 3 and anode 4. The picture of electrolyte flow, speed and pressure distribution in near-surface layers of the deposition area making possible to select optimal geometric parameters of units 2 and 3, have been received due to the earlier performed computer modeling of electrolyte hydrodynamics.

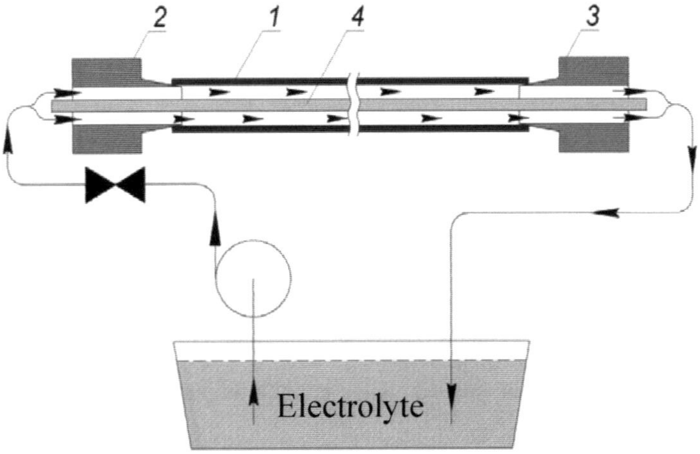

1 – Article to be plated, 2 – Header, 3 – Drain unit, 4 – Anode

Figure 1. Diagram of flowing chrome-plating setup of inner surface of long-size articles

Experimental specimens (Figure 1) were obtained in the course of experimental investigations and the following modifying process parameters: electrolyte temperature (of 50 to 70°C), frequency rotation of article (of 50 to 500 rev/min), working current density being equal to 50 A/dm^2 [3], were selected. Cylinder parts of submerged type sucker rod pumps (GSHN) GSHN RH 1/16 d=27 mm, 700 mm long, chrome plating deposited on their inner surface have been used as articles for investigations. Specimens 100 mm in length were cut out of different parts of articles to carry out investigations.

IOP Conf. Series: Materials Science and Engineering **126** (2016) 012002 doi:10.1088/1757-899X/126/1/012002

Figure 2. Chrome plating specimens

Results

The following parameters of specimens coatings: channel and point porosity, surface micro-hardness of coating, thickness of coatings and its distribution along the article forming have been investigated. The coatings have been studied with the SEM method.

Porosity was determined under GOST 9.302 – 88 «Inorganic Metal and Nonmetallic Coatings. Methods of Control» by superimposing a paper filter. Sodium chloride solution was used for controlling. Average number of pores was calculated as the number of pores along the total specimen surface divided into the specimen surface area (37 cm^2).

Due to the use of electrolyte with UDD and article rotation in the course of chrome plating we managed to get deposits practically without pores average porosity being – 0.063 pores/cm^2.

Micro-hardness is measured on the specimen forming line 100 mm in length with averaging out of eight measurements in each point of measurement. Altogether, each specimen has been studied in 10 points. Measurement have been performed with the hardness gage Constanta TU. Micro-hardness measurement results showed stable surface micro-hardness along the length of all specimens within the range of 710-760 HV. Distribution of surface micro-hardness of specimens is shown in Figure 3.

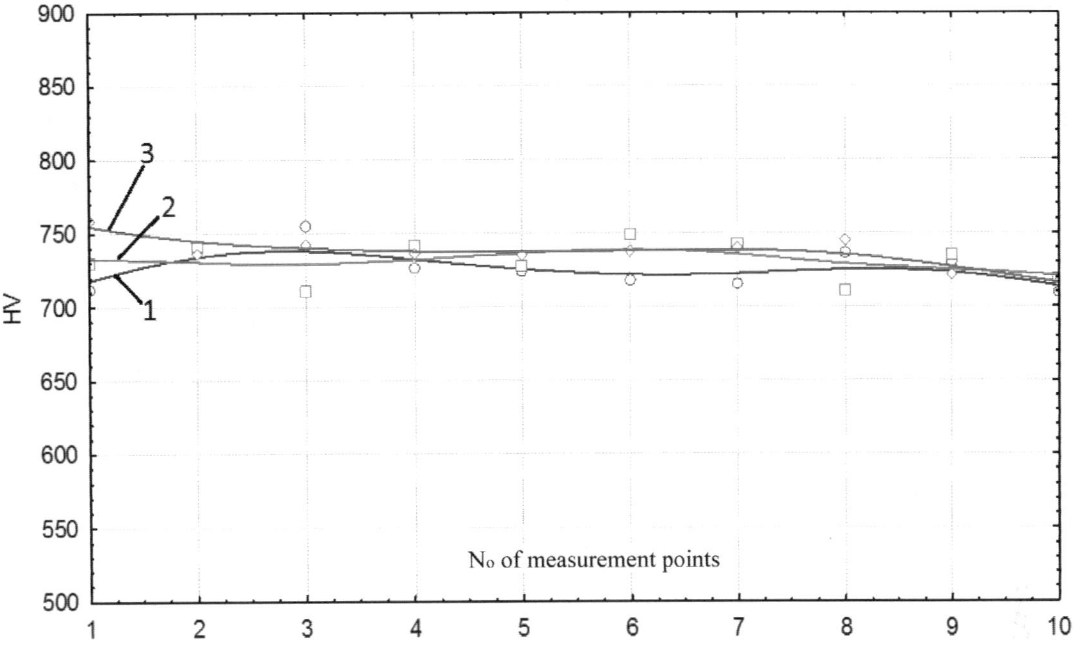

1 – specimens from the zone of feed of the electrolyte; 2 – specimens from the Central zone; 3 – specimens from the zone of discharge of the electrolyte

Figure 3. Distribution of surface micro-hardness of chrome coating.

The coating thickness was determined with the eddy currents method using multipurpose device for measuring geometric parameters, Constanta K6 fitted with ID0 type transducer. The measurement of each specimen has been performed in 10 points along the forming line with averaging in four measurements. The mean thickness of coating in all specimens was 64 μm. The thickness non-uniformity (irregularity) along articles is 5 μm.

The obtained specimens coatings were investigated with the help of scanning electron microscope (SEM) at JSC FR&PC ALTAI. Photographs of coatings and fracture surfaces are presented in Figure 4. Analyzing the image we can draw conclusions. The grains size is getting stable, cracks on the chrome surface on depositing at full rotation of a specimen are less expressed than in case of discrete rotation . Fine-grained structure on the transition boundary chrome-support is clearly.

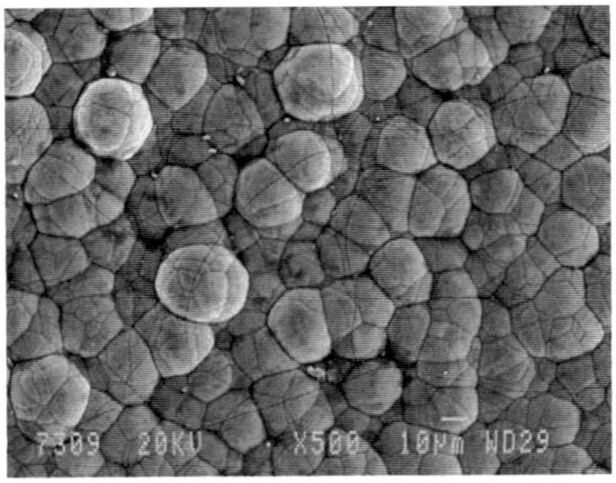

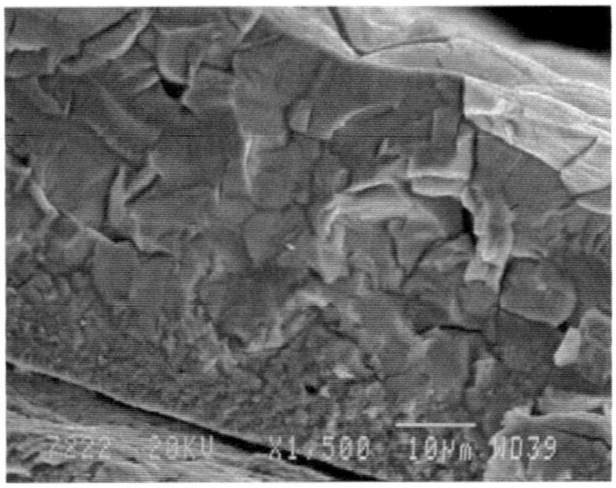

Figure 4. Fracture surface and outside surface chrome coating

Conclusion

1. The structure of chrome plating obtained on full rotation with frequency of 500 rev/min at a temperature 57±1°C is the most fine-grained (reduction in size of coating sub-grains), homogeneous without visible pores. On comparing the coatings obtained at different rotation frequencies, the reduction in sub-grains size of chrome coating vs the growth of rotation frequency can be seen. The grains size is getting stable. The number of dendrites on coating surface also reduces with the growth of rotation frequency.

2. Cracks on the chrome surface on depositing at full rotation of a specimen are less expressed than in case of discrete rotation that testifies less internal stresses.

3. The structure of coating change in thickness is observed on analyzing the fracture surfaces of specimens with chrome plating. Fine-grained structure on the transition boundary chrome-support is clearly seen on specimens. The thickness of fine-grained chrome is of 10 to 15 μm.

Application of electrolytes with UDD for flowing chrome plating of long-sized articles is very promising for obtaining wear-resistant composite electro-chemical coating that are widely used at present, including for special-purpose articles.

References

[1] Azhogin F.F., Belenkiy M.A., Gall I.E., Electroplating. Handbook. Publ. – M.: Metallurgy, 1987. – pp 210-239.
[2] Bogorad L.Ya. Chrome plating. – M.: Mechanical Engineering, 1984. – p 97.
[3] Solodkova L.N., Kudryavcev V.N. Electrolytic chromium plating. – M.: Globus, 2007. – p.191.
[4] Yampolskiy A. M., Ilyin V.A. Short reference book galvanotechnics. – M.: Mechanical, 1981. – p.244.
[5] Koslyuk A.Yu. Investigation of the flowing chrome plating process with UDD. HEMs: Demilitarization, Antiterrorism and Civil Application: Abstracts X International Workshop HEMs-2014. (Biysk, 3-5 Sept., 2014). – Biysk: Publ. Alt. State Tech. Univ., 2014. – pp. 173-176.
[6] Koslyuk A.Yu. Numerical modelling of the current of electrolyte at flowing chromium plating of internal surfaces of lengthy products / Actual problems in machine building: Abstracts I International Workshop. – Novosibirsk, 2014. – pp.38–41.

The Geometrical Model of Calculation of Metal Removal and Roughness in Magnetical-Abrasive Machining

Y Tatarkin, V Xomenko, A Ikonnikova, T Golovina

656038, Russia, Barnaul, Lenin avenue 46, I.I.Polzunov AltSTU

e-mail: aiamagtu@mail.ru

Abstract. The geometrical model of interaction between abrasive particle and the surface in magnetic-abrasive machining is offered. The theoretical research of metal removal is made. The dependences for determination of metal removal and maximum depth of particle introduction are received.

Introduction. The increasing quality requirements for manufactured products cause need for improving existing technologies and creating new ones of finishing with abrasive on a flexible binder (machining with free abrasive condensed with inertial forces, jet-shock, vibro-abrasive machining and a number of other types). The most productive way of abrasive machining on the flexible binder is magnetical-abrasive machining (MAM). The metal removal, here with, is made with the cutting tool created by magnetic field of magnetic-abrasive powder. The cutting tool is defined by increased elasticity [1, 4, 5, 6]. By means of MAM it is possible not only to provide low roughness and polish different forms of details surfaces, but also mechanize such operations, as sharp edges rounding and deburring, descaling and thin oxidic films removing.

Thus there take place reinforcing and surface hardening increase.

Theory. While researching the main regularities of formation of surfaces that are machined with abrasives, one of the major questions is the question of theoretical modeling of interaction process between abrasive particles and workpiece surface.

In the considered model MAM process is described by the following scheme: an abrasive particle having specific size r (circumscribed circle radius) and moving on the surface, inculcates into the material h depth and takes off chips La length (fig. 1).

While calculating next assumptions are made: powder grains have a ball form of one size; working gap is filled with the powder-like environment of density and magnetic properties homogeneity.

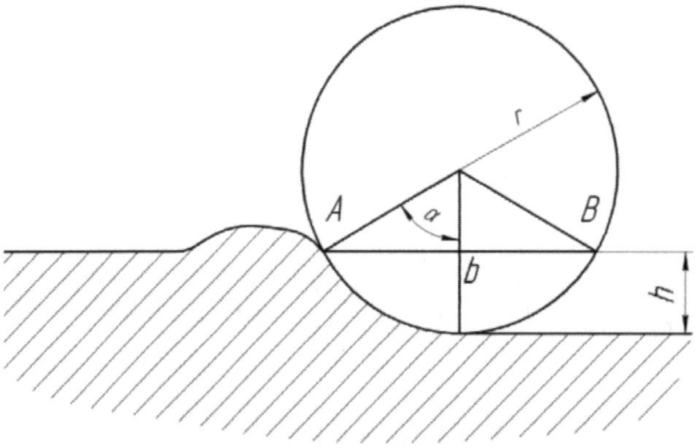

Fig.1 The scheme of single grain material removal in MAM

On the basis of the adopted design scheme and the above assumptions, the amount of material removed

$$V_a = k_c S * L_a = k_c \left[r^2 \arccos\left(\frac{r-h}{r}\right) - (r-h)\sqrt{h(2r-h)} \right] L_a \quad (1)$$

where k_c –chips coefficient, which is equal to the actual area of removed metal considering elastic-plastic deformation of machined material to geometric area, L_a chips length.
The depth of grain forcing into the metal [2, 3, 6]

$$h = r - \sqrt{r^2 - \frac{F_n}{\pi H_v}}, \quad (2)$$

where F_n -normal force, H_v - BHN.
Normal force is equal to P magnetic pressure multiplied by the grain cross sectional area:

$$F_n = p\text{P}r^2. \quad (3)$$

La chip length in mm is determined by n inductor rotation frequency (rpm), $t_п$ (c) polishing time and the width of inductor and workpiece contact zone.
Supposing that the magnets are disposed along the generatrix of the inductor, and the width of contact zone is equal to B core width (Fig. 3), the total chip length for one grain is determined by:

$$L_a = \frac{nB}{60} t_П, \quad (4)$$

where n -the frequency of rotation(rev / min).

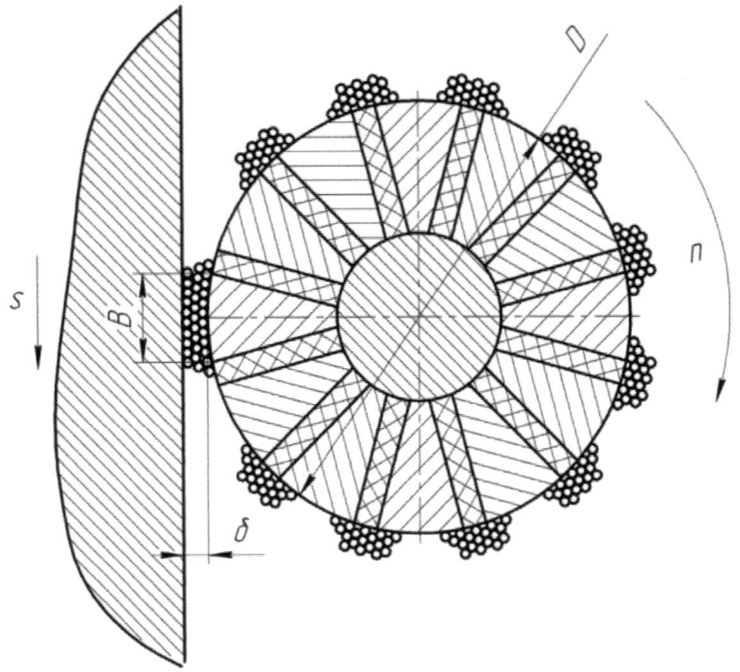

Fig.3 The scheme of workpiece machining.

To determine the amount of grains involved in cutting we denote the diameter of magnetic abrasive particles 3_a, and core area $F_p = B \cdot H \cdot n_M$,

where n_M – core number.

Then the number of working magnetic-abrasive particles is determined by the formula:

$$N_M = \frac{4BHn_M}{\pi \cdot 3_a^2}. \quad (5)$$

Supposing, that n_a cutting abrasive particles are attached to one magnetic particle we get the total amount of cutting grains:

$$N_M = \frac{4BH}{\pi \cdot 3_a^2} n_a n_M \quad . \quad (6)$$

The total amount of removed material

$$V_M = N_M * V_a . \quad (7)$$

To consider the dynamic phenomena during machining we offer the additional coefficient k_d.

It allows to consider the deformation in the tool and workpiece contact zone.

This is due to the fact that during the extensive flat surface machining cutting force causes a considerable deformation of both the tool and the magnetic-abrasive powder.

And the metal removal is reduced.

The coefficient $k_Д$ is always less than 1, but it approaches 1 when the surface significantly smaller than core width B is processed.

This coefficient depends on a large number of factors.

Therefore for it empiric dependence is offered.

Over time, the cutting properties of grains go down because of their bluntness and destruction.

So, we propose to use one more empirical coefficient $k_I(\tau)$ in the formula .

To consider the process of blunting the following empirical dependence is used.

$$k_I = e^{-C_I \tau} \tag{8}$$

Where C_I considers the change of powder cutting properties in time; τ - the total powder working time (c).

If, for example, during τ = 120 c time the powder cutting properties reduce to 20%, then

$$k_I = 0.8 = e^{-120 C_I}. \tag{9}$$

Hence, C_I = 0,00186.

So,

$$V_M = \frac{B^2 H n k_c k_d k_I t_P n_a n_M}{15\pi \cdot 3_a^2}\left[r^2 \arccos\left(\frac{r-h}{r}\right) - (r-h)\sqrt{h(2r-h)}\right] \tag{10}$$

After preliminary cutting marks on the prepared surface look like Fig.3.

Since the profile is periodic, the surface roughness parameters are actually determined by the area $x \in [0; S_z/2]$.

We assume that in this area the mark profile is set by the equation $y = ax^b$.

Then the surface roughness parameters is determined by coefficients a and b. The mark height corresponds to R_{max} parameter.

Herewith

$$R_{max} = a(S_z/2)^b. \tag{11}$$

Ra parameter is determined by the formula

$$R_a = \frac{2}{S_z}\int_0^{S_z/2} |y - W| dx, \tag{12}$$

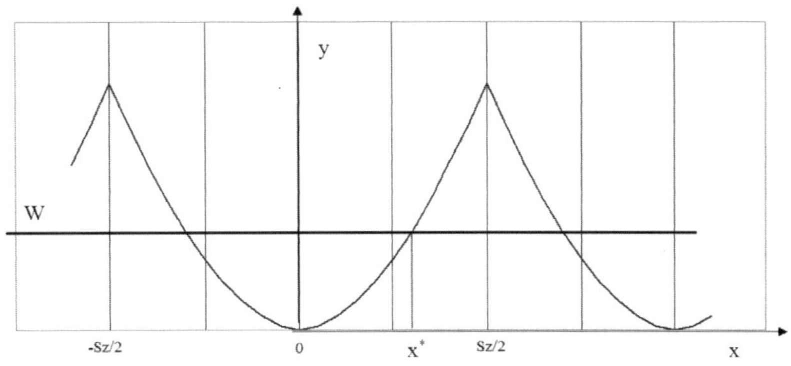

Figure 3 – The shape roughness after cutting.

During processing from workpiece surface having the profile of Fig.3, allowance is removed.
Herewith, the volume of the material that is removed can be calculated taking into account the expression (9).
Metal is removed from mark tops.
Herewith, after some time they take the form shown in Fig. 4.

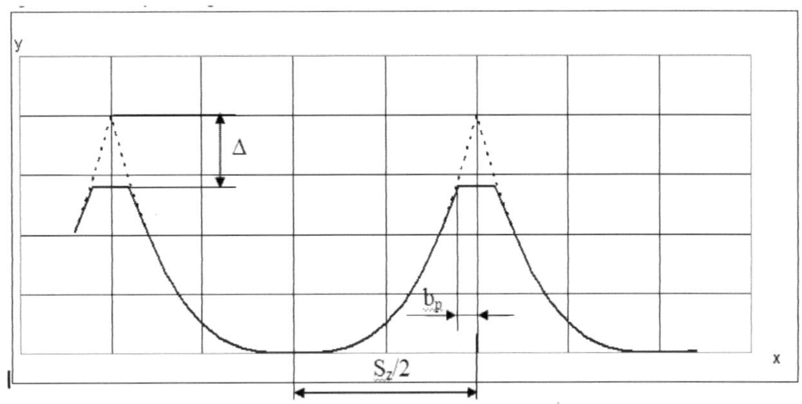

Figure4– Mark profile during machining.

Since the profile of original mark is described by degree dependence, parameters S_zи b_p are related by:

$$\Delta = a\left[\left(\frac{S_z}{2}\right)^b - \left(\frac{S_z}{2} - b_p\right)^b\right]$$

or (13)

$$b_p = \frac{S_z}{2} - \left[\left(\frac{S_z}{2}\right)^b - \frac{\Delta}{a}\right]^{1/b}$$

Dimensions Δ and b_p are determined by metal removal process.

The inductor will go the $b_p$distance for $t_1 = \frac{b_p}{Vp}$ time, where V_p – the feed rate in mm/ s.

Since V_p = S n/60, then

$$t_1 = \frac{60b_P}{Sn}, \tag{14}$$

where S - feed in mm / rev, n - inductor rotational speed in rev /min.
The volume of removed metal in the first pass

$$V_{M1} = \int_{S_Z/2-b_P}^{S_Z/2} ax^b dx - (R_{\max} - \Delta)b_P =$$
$$= \frac{a}{b+1}\left[\left(\frac{S_Z}{2}\right)^{b+1} - \left(\frac{S_Z}{2} - b_P\right)^{b+1}\right] - (R_{\max} - \Delta)b_P \tag{15}$$

On the other hand, this amount is determined by dependence(10) t_p, calculated with the formula (15):

$$\frac{B^2 Hnk_c k_d k_I n_a n_M}{15\pi D_M^2}\left[r^2 \arccos\left(\frac{r-h}{r}\right) - (r-h)\sqrt{h(2r-h)}\right]t_1 =$$
$$= \frac{a}{b+1}\left[\left(\frac{S_Z}{2}\right)^{b+1} - \left(\frac{S_Z}{2} - b_P\right)^{b+1}\right] - (R_{\max} - \Delta)b_P$$

or

$$\frac{4B^2 Hk_c k_d k_I n_a n_M}{\pi S D_M^2}\left[r^2 \arccos\left(\frac{r-h}{r}\right) - (r-h)\sqrt{h(2r-h)}\right]b_P - \tag{16}$$
$$- \frac{a}{b+1}\left[\left(\frac{S_Z}{2}\right)^{b+1} - \left(\frac{S_Z}{2} - b_P\right)^{b+1}\right] + (R_{\max} - \Delta)b_P = 0$$

where Δ is calculated with the formula (16).

Equation (14) can be solved for b_P only numerically, excluding trivial solution $b_P = 0$.

Fig. 5showsthe dynamics of roughness parameter changes Ra from pass to pass when feeding S = 0.05mm / rev.

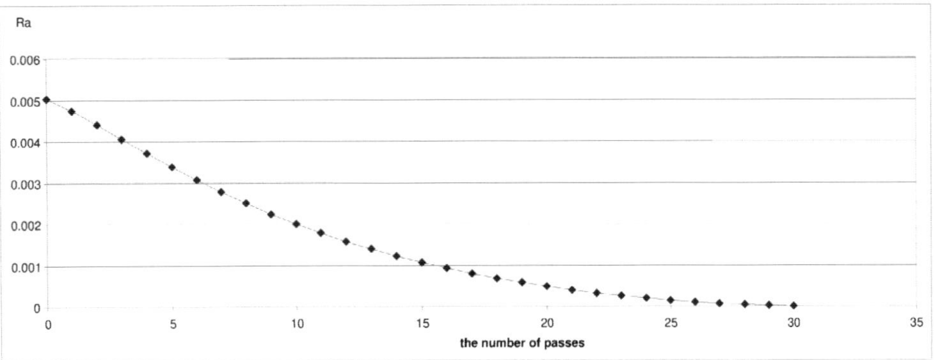

Figure 5 - The dynamics of roughness parameter change Ra

The program for performing calculations is designed in VBA in MS Excel.
The calculation time on modern computers is less than 1 second.

Results and discussion. The proposed program allows to calculate the metal removal and the surface roughness in magnetic-abrasive machining, depending on the chosen parameter of machining.

Conclusions. The dependences derived from modeling can help to accomplish some vital engineering tasks dealing with the workpiece finishing process. They can be used to determine optimal sizes of abrasive powder portions participating in the process of magnetic abrasive machining, an appropriate technical rate setting, the most effective trajectory and operation modes.

Using the obtained results a production manager can assess a potential effect of some process related parameters (e.g. an inductor radius, the radius of the powder portion, the machine table feed, the rate of the inductor rotation, etc.) on the production efficiency. The model presented above is of great scientific value because it can be used for parameter prediction for machining sophisticated flat workpieces when machining conditions are not constant and have to be readjusted due to the peculiarities of technical characteristics of the magnetic abrasive machining process.

References

[1] Baron Y.M. Magnetic-abrasive and magnetic machining of products and cutting tools. - L .: Mechanical engineering. Leningr. Department, 1986.-176 p.

[2] Ikonnikov A.M, Fedorov V.A The parameters calculation of magnetic-abrasive machining of complex profile surfaces. "Metal processing ", 2003, №4. Novosibirsk, OJSC STE and EI "ORGSTANKINPROM." - P.10-11.

[3] Y. S. Stepanov, G. V. Barsukov, A. V. Mikheev, Modeling of the number of abrasive grains that are involved in the process of forming the surface of a part during supersonic water jet stream cutting. News of Brjansk State Technical University. 1 (2009). 10.7862/rm.2012.11.

[4] A. G. Kolesov, S. E. Sazonov, S. M. Bratan, Development of a mathematical model of the polishing process for machining precision parts of devices. News of Sevastopol National Technical University. 139 (2013) 111-116. 10.15276/etks.14.90.2014.16.

[5] F.Y. Sakulevich, Fundamentals of magnetic abrasive machining. Science and Engineering, Minsk, (1981). 10.4028/0-87849-974-1.281.

[6] Jayswal, S.C., Jain, V.K., Dixit, P.M., 2005b. Modeling and simulation of magnetic abrasive finishing. International Journal of Advanced Manufacturing Technology 26, 477–490.

Finite Element Surface Layer Inheritable Condition Residual Stresses Model in Surface Plastic Deformation Processes

M S Mahalov [a], V Yu Blumenstein [b]

T.F. Gorbachev Kuzbass State Technical University, 28, street Vesennyaya, Kemerovo, 650000, Russian Federation

[a] e-mail: maxim_ste@mail.ru, [b] e-mail: blumenstein.vu@gmail.com

Abstract. The residual stresses (RS) research and computational algorithms creation in complex types of loading on the product lifecycle stages relevance is shown. The RS forming finite element model at surface plastic deformation strengthening machining, including technological inheritance effect, is presented. A model feature is the production previous stages obtained transformation properties consideration, as well as these properties evolution during metal particles displacement through the deformation space in the present loading step.

Introduction

It's known that the first kind residual stresses (RS) - one of important metal condition parameters, which largely determines machines critical parts and industrial constructions operational life. Especially RS great impact on the durability in different types of alternating loads, the RS value and parts section distribution have important role. In case of workpiece surface stress concentrator or initial crack appearance the compressive RS positive effect in the surface layer (SL) increases sharply and can give the several times parts durability increase [1].

The main condition for the first kind RS appearance is uneven over the workpiece cross section thermal or mechanical plastic deformation. However, due to the complexity of the parts power and thermal load processes in production and operation stages, accurate stresses calculation, acting in a particular part or parts of industrial construction, it is not always possible [2].

Most computational methods for stress-strain condition (SSC) determining is based on the large numbers of assumptions, idealizing computational schemes, loads and boundary conditions. It results to significant errors in the stresses evaluation, and in some cases makes impossible to obtain the theoretical solutions.

Such problem is becoming especially actual when you need to define RS in the product machining processes and their influence at subsequent operational stage processes. RS calculation methods at each machining stage has to consider surface layer complex nonmonotonic loading, metal plastic flow in DZ and deformation parameters technological inheritance (TI).

Theory

Among the machining methods high potential for generating in the SL of positive compressive RS are similar to the metal yield stress values have ways of finish-strengthening processing by surface plastic deformation (SPD) (Fig. 1) [3].

During SPD treatment in the deforming tool and workpiece contact area occurs asymmetrical deformation zone (DZ) ABCDEFG, limited contour lines front non-contact ABC (zone 1), front contact CD (zone 2), back contact DE (zone 3) and a rear non-contact EF (zone 4) surfaces and also FGA curve, describing the boundary of metal plastic flow zone [4]. Due to the stress and strain fields impact, metal particles are displaced in the area of wave along the some metal plastic flow lines, forming a workpiece SL (Fig. 1). The initial condition parameters, which the metal particles have before enter the deformation zone (line AG), are transformed into the accumulated on DZ exit (line GF).

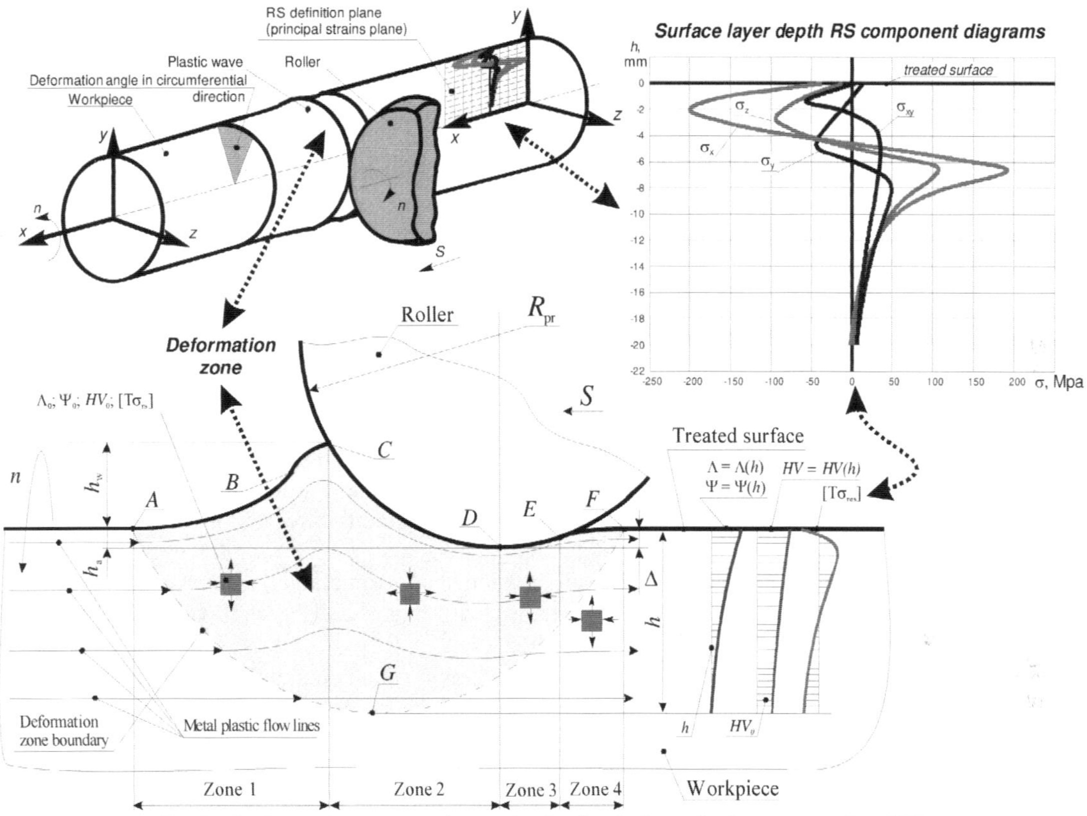

Fig. 1. Deformation zone scheme in the feed plane during processing SPD

Solution of the analytical RS determination after complex loading types taking into account TI effects is possible through device deformed body mechanics and methodology TI mechanics instruments in which:

- the SL condition formation and transformation in the machining stages and subsequent operational loading are considered as a uniform continuous process of strain accumulation, plasticity reserve exhaustion and RS transformation in surface layer metal. Thus, along with traditional quality parameters, for describing SL properties respectively using: the cumulative shear strain level Λ, the plasticity reserve exhaustion level Ψ and residual stresses components tensor $[T\sigma_{rs}]$ in coordinate system, associated with the workpiece or construction element form;
- the residual stresses formation and transformation in the manufacturing machining steps considered as complex non-monotonic metal loading result in DZ, in which the surface layer strain accumulation and metal plasticity reserve exhaustion occurs;
- the RS formation occurs in the metal mechanical properties changing due to plastic deformation conditions: loading history affects to the processes flow at every trajectory point inside DZ;

- final RS manifest themselves during the operation, transforming in each operating loading cycle; durability defines entire load prehistory.

As is known, RS tensor may be represented by the expression

$$[T\sigma_{rs}] = [T\sigma_{rep}] - [T\sigma_{iel}] + [T\sigma_{uwr}] + [T\sigma_{t}], \quad (1)$$

where components are [5-6]:
$[T\sigma_{rep}]$ – stresses tensor, appearing in the real elastic-plastic body under loading;
$[T\sigma_{iel}]$ – stress tensor, which would occur in a ideal elastic body under identical loading;
$[T\sigma_{uwr}]$ – with unload elastic stresses tensor when workpiece releases;
$[T\sigma_{t}]$ – unload elastic thermal stresses tensor.

According to the TI mechanics instrument:

- at each machining step in DZ (as described above on SPD processing example) is a continuous strain accumulation and plasticity reserve exhaustion occurs, which leads to the surface layer formation with defined quality parameters: the hardening depth and degree, the roughness and the residual stress;

- in the operational fatigue loading, the strain accumulation and plasticity reserve exhaustion processes continue, flowing under continuous residual stresses relaxation. With the strain limit accumulation ($\Lambda = \Lambda_p$), there is a complete plasticity reserve exhaustion occurs ($\Psi = 1$). This condition corresponds to the first kind residual stress tensor relaxation to negligible values ($[T\sigma_{rs}] \approx 0$) and the fatigue crack appearance. A strain limit accumulation, a complete plasticity reserve exhaustion and fatigue crack nucleation occurs at a probable failure point, which may be located both on the surface and at some distance from it. The resulting stress condition determined by residual and external load stresses tensors. When the both constituents principal stresses vectors direction match and a fatigue stresses cycle is symmetrical, the workpiece (machine part) surface layer works in a cycle asymmetry condition all the more, than the residual stresses greater. The average cycle stress is equal to the residual stress, and the amplitude is equal to the external load stress. Under cyclic loading compressive residual stress increases and tensile - reduce fatigue life;

- at each machining or operation step the residual stress forming occurs under influence of:
  - the strain level, metal accumulated by metal at the current time;
  - the residual stresses, generated in the previous load step;
  - the mechanical and thermal stresses, generated during a load application in the present load step;
  - the unload elastic stresses, when the load is removed;
  - the stresses, additionally appearing when workpiece releases.

Hereditary SPD Process and Surface Layer Condition Parameters Forming Model

For a complete TI phenomenon account and its influence on the SL metal final properties formation necessary to create a theoretical model representations about the metal stress-strain condition is not only taking into account the properties transformation, obtained in previous production and operation stages, but also taking into account changes in these properties during the metal particles move through DZ space at the current load stage.

At the same time, if we consider the principal strains plane (Fig. 1), the feed direction tool and DZ displacement has a discrete character: next DZ appears with offset from the previous on the feed value, which upon SPD treatment usually varies between 0.05-0.3 mm/rotation.

In the early finite element (FE) SPD, cutting and dimensional joint running-in process models, proposed in [3, 7], said discrete not modeled on the one hand simplifies the model and subsequent calculations, on the other hand - it is a disadvantage.

In the previously proposed models, the workpiece upper boundary was obtained experimentally a deformation zone contour. The contact area with the roller was simulated as an arc, which is identical to the roller profile radius (Fig. 1). The model lower boundary was rigidly fixed at both axes, roller has a single offset along the x-axis with feed value S. This allowed to simulate the «instantaneous» stresses and strains values, that occur in DZ space.

A feature of this model is that the tool doesn't interact with the entire treated surface - simulated only SSC in DZ, whose geometry is obtained on the basis of experimental data. It's assumed that the metal particles pass through DZ and form a treated SL. Moving through DZ space along some flow-line, due to the stress and strain fields, metal particle has a complex non-monotonic loading. To calculate the accumulated mechanical condition parameters was used the hypothesis, that in the next load step, which occurs via 1 workpiece rotation, the particle, moving along the flow-line, changes its SSC on some new, located in DZ space from the previous along workpiece axis with offset on the feed value. Wherein it doesn't take into account that between these two conditions occurs unload, leading to an intermediate stress-strain condition appearance, and a next load step occurs in the metal properties conditions, have changed in the previous step.

In this paper, for SL metal hardening treatment by roller tool SPD was created the RS forming finite element model, taking into account inside DZ metal properties transformation and also integral metal mechanical condition parameters calculated. At the same time we made the following initial conditions and assumptions:

1. The occuring strain had only mechanical nature: the temperature component is not taken into account because of the metal small heating in the loading process.

2. Simulated isotropic material with the mechanical condition parameters corresponding to parameters of steel 45 (GOST 1050-88, hardness 160-180 HV) in the delivery state.

3. Flow curve linear approximation, taking into account the metal hardening (Fig. 2), is used. The following parameters were used for description:

- Young's modulus E, which characterizes the elastic section angle α;

- extrapolated yield stress σ_{eys}, corresponding to stresses axis segment, which is determined by the plastic flow line extension;

- Poisson's ratioν;

- tangential module T_{mod},, which characterizes the plastic flow line angle β.

4. In view of the relatively low error value, to simplify the calculations, don't include the Baushinger effect.

5. The workpiece releases unload elastic stresses tensors was adopted zero because a set of conditions, specified in [8] for the symmetric cylindrical workpiece treatment, was performed.

6. Appearing edge effects don't significantly affect the simulation results, because workpiece model was large enough compared to the appearing deformation zone.

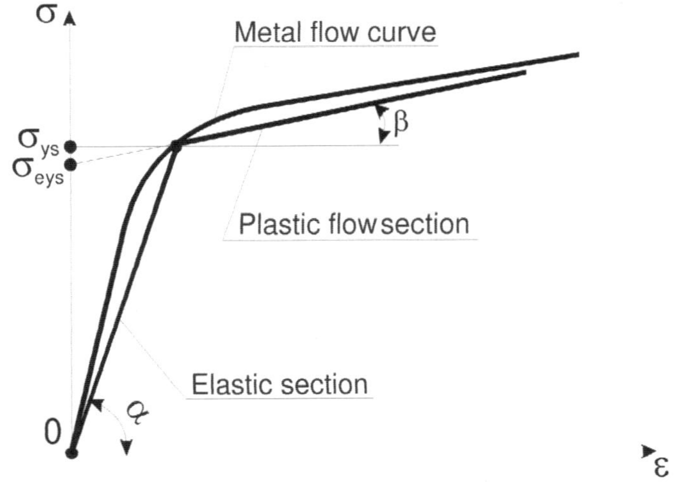

Fig. 2. Bilinear flow curve approximation diagram

The problem was solved in the plane-strain formulation. It is assumed that all deformation processes in DZ occurring in the principal strains plane, passing through the workpiece rotation axis (feed plane). This is confirmed by SPD processes research results, presented in [4], according to which:

- plastic wave height (ABC, Fig. 1) in the longitudinal direction is much larger than in a cross;
- feed plane largely corresponds to the main strain plane concept, since the rotation speed plane strain is less than a similar in feed plane.

Belonging to principal strains plane, treated cylindrical workpiece fragment of length L = 50 mm and height h = 20 mm was simulated (Fig. 3).

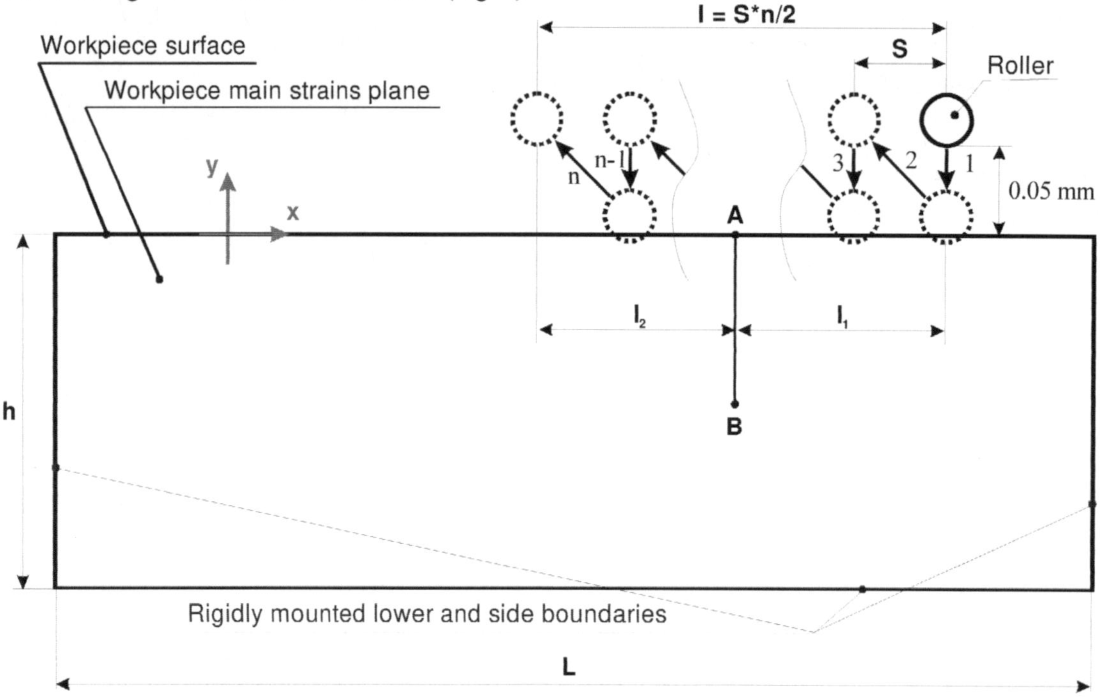

Fig. 3. SPD processing simulation problem statement scheme

The fragment lower and side boundaries rigidly fixed on both axes. Deforming roller simulated as an absolutely rigid body, having a determined profile radius arc.

This assumption is justified, because numerous researches have shown, that during the SPD treatment, in the surface interaction process:

- deforming tool has very small abrasion;
- any significant tool deformation is not fixed.

During the simulation, the selected SPD treatment settings located approximately in the middle of used ranges (Table 1).

Table 1. Simulated SPD treatment parameters settings

Parameter mode	Symbol	Measure unit	Simulation value	Parameters value range
Roller profile radius	R_{pr}	[mm]	5	1.6-16
Feed	S	[mm/rot]	0.1	0.05-0.25
Actual roller tightness	h_a	[mm]	0.05	up to 0.1 mm for steel 45

In the initial position the tool has clearance of 0.05 mm from the surface (Fig. 3).

On the first and any subsequent odd simulation step was performed loading - the surface direction tool displacement with the value of 0.1 mm. Thus, the tool being indented into the surface with the actual tightness value.

On the second and any subsequent even simulation step was performed unloading - the surface direction tool retraction with the value of 0.1 mm, while moving along the surface with the feed value. Thus, the tool again had a relative to the surface initial clearance.

A total 300 steps was simulated - 150 loading steps and 150 unloading steps. The treated surface length in the presented model was

$$l = S*n/2 = 15 \text{ mm}. \quad (2)$$

The basic modeling idea of the stress-strain condition and residual stresses transformation in deformation zone is as follows. Certain modeled fragment section AB is positioned so, that at the first load step is not yet in the DZ space. At the same time at the penultimate load step this section has already left DZ space. In the feed direction distance l_1 (from the tool starting position to the section AB) was 5 mm, distance l_2 (from section AB until the end tool position) - 10 mm (Fig. 3-4).

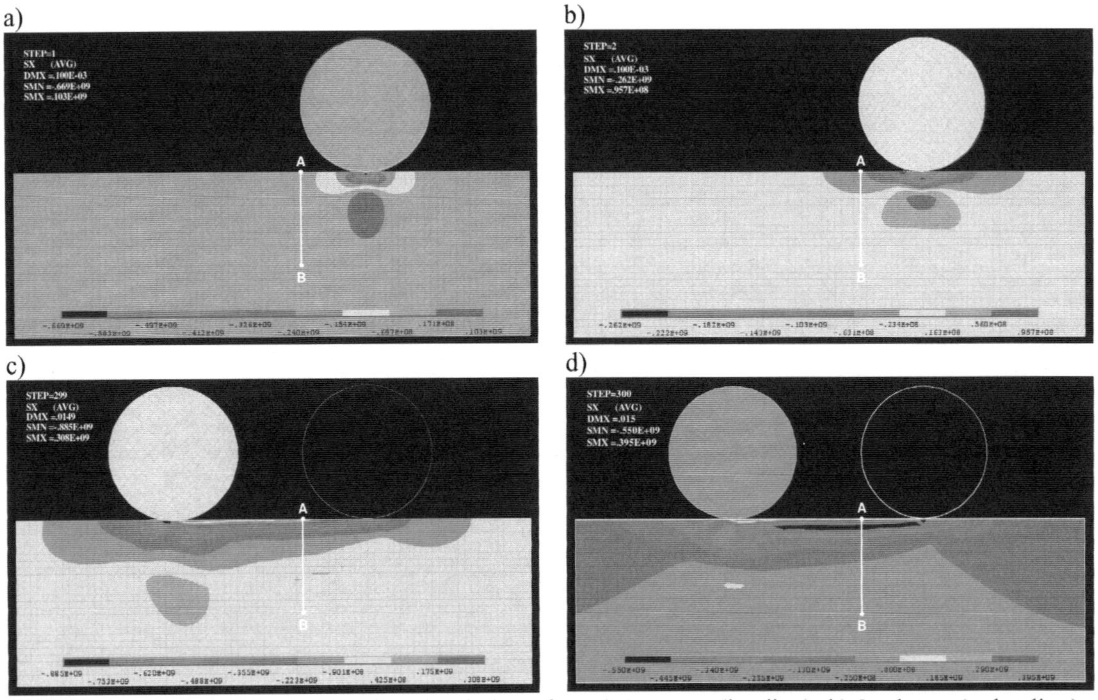

Fig. 4. The axial stresses distribution simulation after: a) 1-st step (loading); b) 2-nd step (unloading); c) 299-th step (loading); d) 300-th step (unloading)

As, in the all modeling steps implementation said section passes through a DZ space, this section is the treated SL depth trajectory, in which strain accumulation, partial plasticity reserve exhaustion and RS tensor forming based on changing the SL properties was happened. Fig. 4 is an example of the axial stresses distribution after the 1-st, 2-nd, 299-th and 300-th load steps (marked section in which SSC parameters at each simulation step are fixed).

For further mechanical condition parameters calculations, after the model solution receiving in a selected section AB for each simulation step was fixed the following parameters: the nodes coordinates, the nodes displacement vector components, the stress tensor components, the total elastic-plastic strain tensor components. The following also have been taken in the calculation:

- workpiece rotation frequency 300 rot./min, then, one workpiece rotation time is 0.2 s;

- the circumferential direction deformation angle (DZ space) 10° (taken from the data of [3-4]), then, one SL loading and unloading cycle time is 0.0054 s.
- half of this time (0.0027 s) loading occurs, and as many - SL unloading.

Results and Discussion

Mark some RS tensor SL depth distribution features (Fig. 5):

1. The largest compressive stress values (up to 560 MPa) has an axial component σ_x. These values are close to the hardened metal yield stress, which is favorable because of a significant influence of this component to the workpiece (machine part) cyclic durability during it operational (fatigue) loading. Compressive stresses σ_x on the surface and at a nearby layer, summed with the operating tensile stresses, cancel each other, thereby reducing the acting stress tensor.

These values correspond to the experimental data, obtained in [4], where the axial RS value at SPD reaches -800 MPa, and the circumferential RS (σ_z in Fig. 5) - up to -400 MPa.

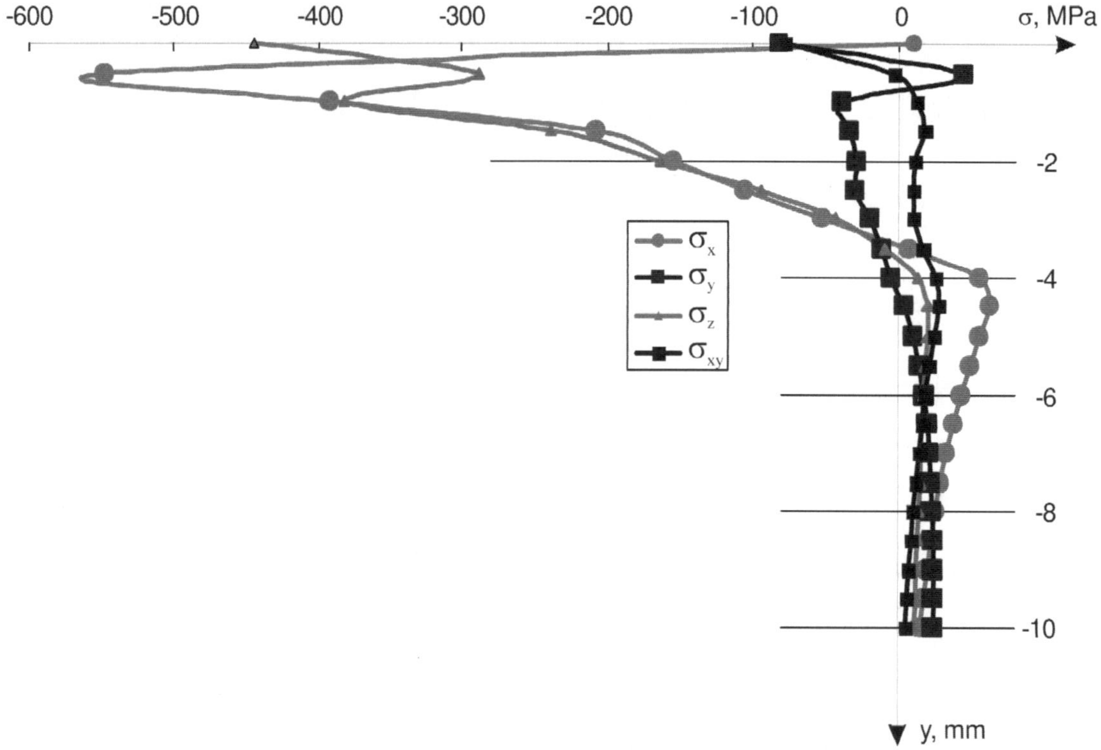

Fig. 5. The residual stress tensor components distribution in the surface layer depth after the 300-th simulation step (y = 0 mm corresponds to the workpiece surface)

2. The axial component detects the compressive stresses extremum, located at a depth of 0.5 mm below the workpiece surface. This is consistent with the results of [9], in which the author emphasizes the compressive RS extremum possibility on the workpiece surface, and at a certain depth.

3. The component σ_y (radial RS) has a considerably smaller values than σ_x and the same depth spread. On the surface and in the surface layer σ_y has a predominantly negative values not exceeding -100 MPa. It also has smaller differences as to the region of tensile and compressive values to. This corresponds to the results obtained in [4], according to which the SPD surface radial component has zero value. Wherein the author notes, that the characteristic SPD RS tensor small σ_y values are not significant disadvantage due to insignificant influence of this component to the workpiece cyclic durability.

4. The residual stresses depth distribution reaches up to 7 mm, wherein a significant residual stresses normal components compressive values depth distribution has obtained (up to h = 4 mm).

Overall, the picture is in agreement with the analytical and experimental results obtained by the previously published works authors.

Conclusion

1. The SPD finite element model, was developed. Its feature is the deforming tool with elastic-plastic half-space discrete (in principal strains plane) interaction nature accounting. Discrete interaction accounting achieved by the original scheme of multistep loading, unloading and new deforming tool feed displacement inside the deformation zone.

2. Within each «loading-unloading» cycle the simulation, metal stress-strain condition and residual stresses tensor calculation was performed

3. Each cycle loading and unloading history accounting allowed to estimate the TI phenomenon influence to the surface layer metal properties formation process. TI phenomenon influence estimation is available not only at the life cycle (treatment and operation) stages level, but also inside each stage by taking into account inside deformation zone metal properties changes.

4. The obtained residual stress tensor components numerical values reflect the metal properties formation hereditary nature inside the deformation zone during the surface plastic deformation process.

5. Obtained results allow to perform the subsequent accumulated SL metal mechanical condition parameters calculations, taking into account the TI phenomenon.

References

[1] S.I. Ivanov, V.F. Pavlov, Influence of residual stresses and hardening on fatigue strength, Strength Problems. 5 (1976) 25-27.

[2] M.S. Mahalov, V.Y. Blumenstein, Residual stresses simulation on the products life cycle stages, Mechanical Engineering Journal. 12 (2014) 21-25.

[3] V.Y. Blumenstein, V.M. Smelyanskiy Technological Inheritance Mechanics in the Machine Parts Treatment and Operation Stages, Mechanical engineering-1, Moscow, 2007.

[4] V.M. Smelyanskiy, Machine Parts Surface Plastic Deformation Hardening Mechanics, Mechanical engineering, Moscow, 2002.

[5] I.A. Birger, Residual Stresses, State scientific and technical publishing engineering literature, Moscow,1963.

[6] M.V. Storozhev, E.A. Popov, Metal Forming Theory, Mechanical engineering, Moscow, 1977.

[7] V.Y. Blumenstein, M.S. Mahalov, Hardened surface layer residual stress computational model after surface plastic deformation treatment, Hardening and coating technology. 4 (2013) 12-20.

[8] A.N. Ovseenko M.M. Gajek, V.I. Serebryakov, Formation of the Machine Parts Surface Layer Condition by Technological Methods, Politechnika Opolska, Warsaw, 2001.

[9] D.D. Papshev Machine Parts Running-in Balls-Hardening, Mechanical engineering, Moscow, 1968.

Surface Layer Formation When Finish-Hardening Processing of the Parts by Smoothing

V N Belyaev [1,2,a], E Ju Tatarkin [3,b]

[1] Biysk TechnologyInstitute, 27 Trofimovast., Biysk, 659315, Russian Federation
[2] JC @FRPC "Altai", 1 Socialisticheskayast., Biysk, 659322, Russian Federation
[3] I.I. Polzunov Altai state technical university, 46 Lenina pr.,Barnaul, 656038, Russian Federation

[a] e-mail: tmp-bti@ya.ru, [b] e-mail: etatarkin@mail.ru

Abstract. Problems of surface layer formation of the parts, when hydraulic smoothing, are considered in this work. The results of theoretical and pilot studies of smoothing in case of nanocarbons and copper salts introduction into the process liquid are given. The influence dependences of the processing modes on roughness and microhardness of surface layer are defined.

1. Introduction

Increasing of operating life of machine parts caused the necessity to develop the new methods of processing and to improve the current ones. One of the promising directions is the usage of hardening processing in combination with different influences (chemical, thermal, ultrasonic and etc.) upon the processed surface, and the development of adaptive influences upon it, including the impacts based on basic research in contact interaction of the materials (Rehbinder and Kirkendall effects, selective transfer effect and etc.). One of the examples of such direction realization when SPD is Papshev's studies. He showed that microhardness of the processed surface is improved in case of SAS introduction into coolant (Rehbinder effect).

There is a SPD processing of parts where the effort of deformation is provided by the deforming ball, affecting the detail under pressure of process liquid from hydro-power station – hydraulic smoothing (Fig. 1). The advantages of the given method in comparison with the "classic" smoothing are: lower process liquid flow when supply onto the processing zone and constant effort of deformation onto the processed surface.

Studies show that it is possible to influence on the parts precision when hydraulic smoothing by means of parameters control of process liquid (viscosity, pressure). However, there is practically no data concerning the influence of process liquid characteristics when hydraulic smoothing (chemical properties, presence of fillers, viscosity and etc.) upon the quality parameters of surface layer (roughness, microhardness, wear resistance and etc.).

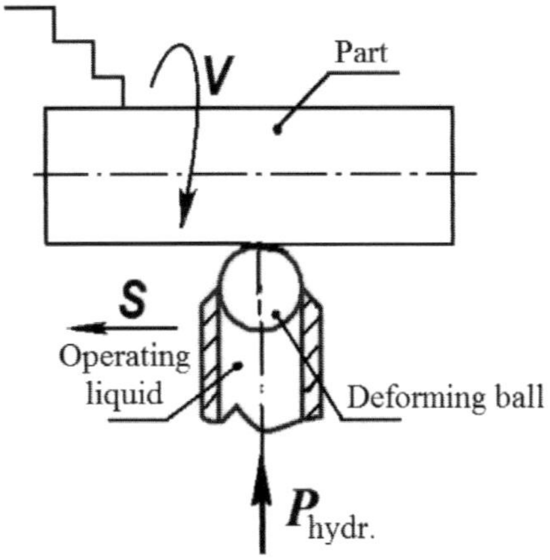

Fig. 1 – Hydraulic smoothing scheme

2. Theory

The quality of surface layer when SPD is formed under the process of metal deformation, and the process itself depends on interaction conditions of the instrument and part: metal strain, friction ratio, process modes and etc. It is well-known that the change of process of frictional contact of friction pair can be obtained by introducing friction modifiers into lubricants, such as organic additives, metal and non-metal powders (graphite, molybdenum disulfide, copper and etc.). The introduction of solid friction modifiers into the deformation zone when SPD increases load-carrying capacity of the lubricant, prevents from "dry friction" in case of great loadings. Throughout the process, impregnation of particles-filler onto the contacting surfaces can be observed. So we can suggest that the introduction of solid particles into process liquid when hydraulic smoothing may provide the change of metal deformation process and introduction of particles-filler onto the article surface, and as a result, the improvement of operating characteristics.

The effect of "selective transfer" is well-known from Garkunov's basic research[1]. This effect allows formation of antifriction coatings on the surfaces of machine parts. The performance of formation the coatings when SPD in case of "classic" supply of coolant is shown in Sorokin' and Bersudskiy's studies and other scientists as well. The analysis made the following supposition: when hydraulic smoothing, it is possible to coat thin copper layers onto the surface of parts if copper salts were been introduced into the process liquid[2,3]. This hypothesis is based on copper reducing reaction on the processed surface, where glycerine serves as electron sources for copper irons reduction (Fig. 2).

Liquid between the deforming ball and processed surface when hydraulic smoothing allows realization of the proposed hypotheses[4]. According to contact-hydrodynamic theory of lubricant, the liquid layer thickness between friction surfaces is the following[5,6]:

$$h = f(\mu_P; \vartheta_\Sigma; \rho), \qquad (1)$$

where μ_0 – liquid viscosity;

$_\Sigma$ – total angular velocity of the blank and rotating deforming ball;

ρ – reduced surface radius.

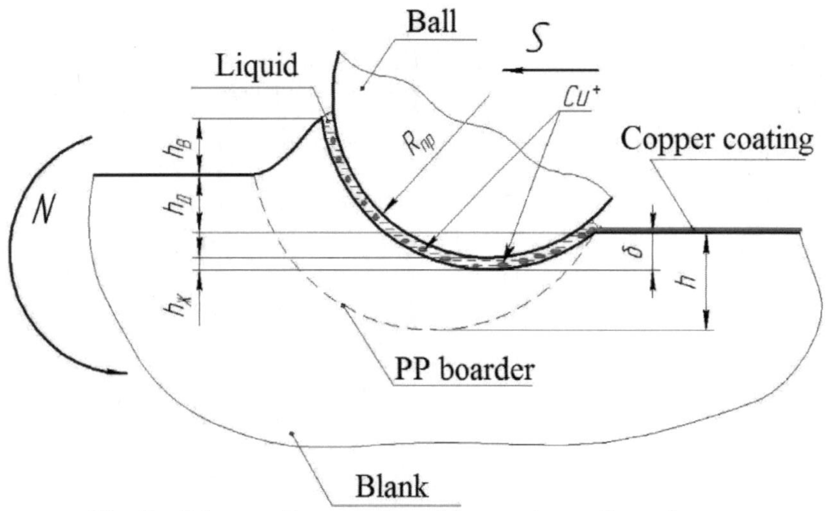

Fig. 2 – Scheme of copper deposition on the surface of a part

Judging by the fact that liquid viscosity depends on the temperature, filler and pressure

$$(\mu_P = \mu_0 e^{\alpha P}) \tag{2}$$

then total angular velocity can be defined as follows:

$$\vartheta_\Sigma = \vartheta_{Bl.} + \vartheta_B = \frac{\pi \cdot n}{1000} \cdot (D_{Bl.} + d_B), \tag{3}$$

and reduced radius is:

$$\rho = \frac{D_{Bl.} \cdot d_B}{2 \cdot (D_{Bl.} + d_B)}, \tag{4}$$

so we may conclude that liquid layer thickness h depends on such process parameters of mechanical processing by smoothing as process liquid viscosity μ, deformation pressure P, processing speed (part rotation frequency) n. Thus, factors, affecting the surface layer formation when hydraulic smoothing, can be presented in the form of diagram (Fig. 3).

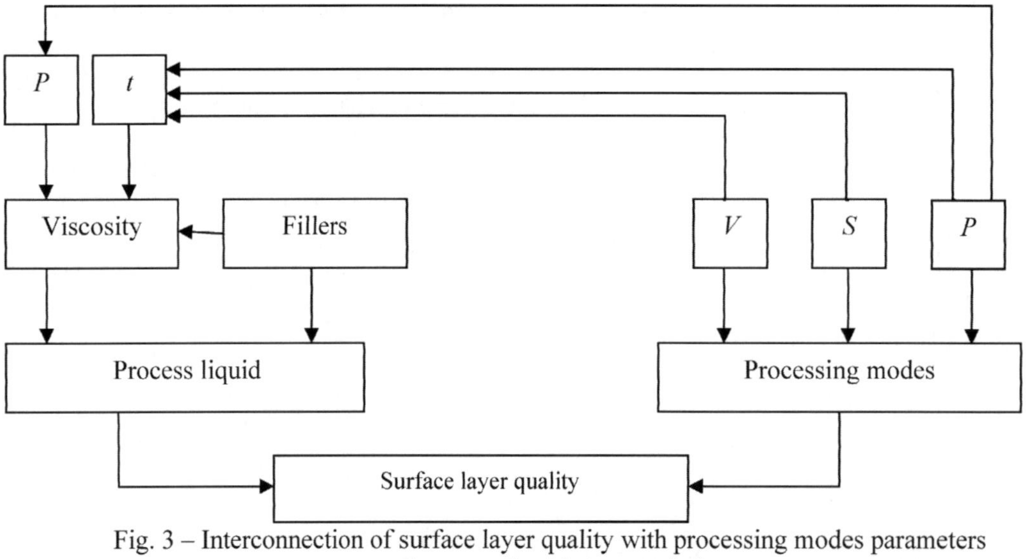

Fig. 3 – Interconnection of surface layer quality with processing modes parameters and process liquid characteristics

3. Results and discussion

Pilot studies of hydraulic smoothing on the screw-cutting lathe 16K20F3 type were conducted to confirm the proposed hypotheses. Tested parameters are roughness *Ra* and microhardness *HV*. Processing parameters are given in Table 1.

Table 1 – Parameters of pilot studies of hydraulic smoothing

Parameters	Performance	
	"Particle impregnation" hypothesis	"Copper film formation" hypothesis
Filler	UDAG-S, Al_2O_3	$CuSO_4*5H_2O$
Filler concentration	0-1 % mass.	25–30 g/l
Process liquid	industrial oil И-40А	glycerine
Tool advance, mm/rot	0.05 - 0.2	0.05 - 0.2
Blank rotation speed, m/min	40 – 165	40 - 165
Pressure in hydraulics, MPa	3 – 12	3-5
Blank material d= 65 mm	steel 45 State Standard (GOST)1050-88	steel 45 State Standard (GOST)1050-88 alloy AL-30

Pilot studies were conducted with samples made of steel 45 GOST 1050-88 and aluminum alloy AL-30. These materials are widely used in engineering- and automobile industries to manufacture articles and parts operating in friction modes and running in before (shafts, ICE pistons and etc.).

Observed dependences *Ra* and *HV*, given in Fig. 4 – 9 were obtained when studying the "particle impregnation" hypothesis.

The obtained dependences show that nanocarbons provide formation of the surface with less roughness in case of surface plastic deformation. The increase of particle concentration in operating liquid leads to the decrease of surface roughness. It can be explained by the decrease of pimples amount due to the diamond-carbon particles (UDAG-S) during flowing of liquid with particles relative to the processed surface. Fig. 4 shows that the etreme concentration is not achieved yet. The increase of roughness should be after it, due to the achievement of so called "dry friction" when low amount of liquid component in deformation zone. Distance between the tops of pimples is lengthened when increasing the tool advance due to the reduction of pressure impacts per unit of deformation area. It results to the decrease of plastic deformation caused by less influence of the deforming ball on the pimples. Pressure elevation of operating liquid in hydraulics within the studied range leads to greater metal deformation and less roughness of the processed surface. Blank rotation speed almost has no influence on roughness (Fig. 6).

Introduction of UDAG-S particles into metal deformation zone causes the surface layer to harden much due to the locating of shear stress in thin layer and increasing of normal stress, providing great compression of pimples and metal hardening.

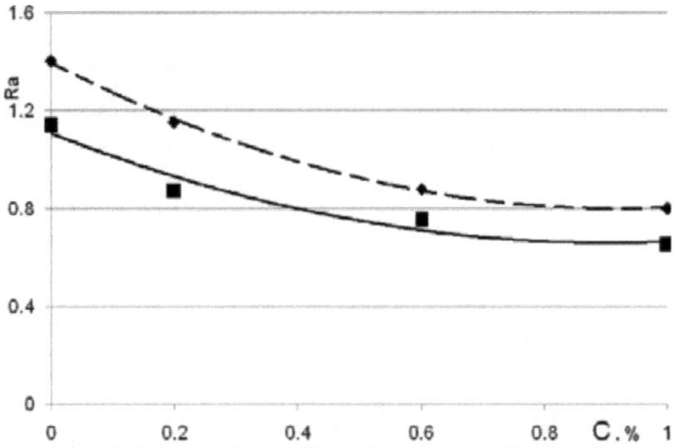

- - - S=0.05 mm/rot., V=40 m/min., P=4 MPa
—— S=0.05 mm/rot., V=40 m/min., P=5 MPa

Fig. 4 – Dependence of roughness on UDAG-S concentration

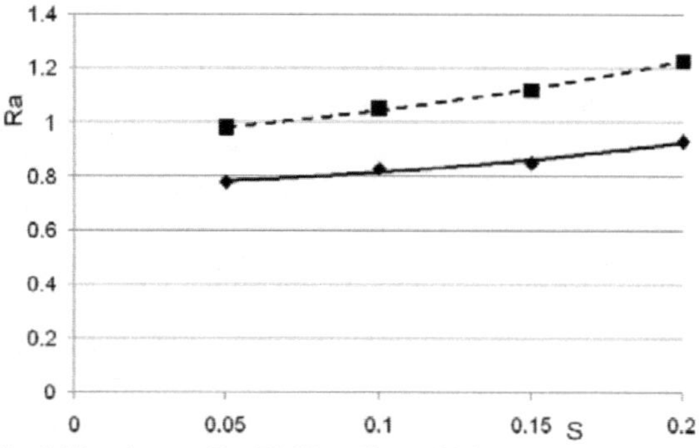

- - V=165 m/min., P=5MPa, C_{NC}=0% mass.
—V=165 m/min., P=5 MPa, C_{NC}=1% mass.

Fig. 5 – Dependence of roughness on tool advance

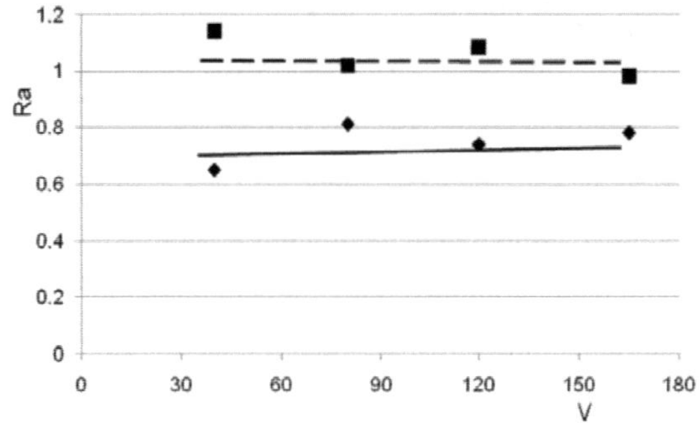

- - - S=0.05 mm/rot., P=5 MPa, C_{NC}=0% mass.
—S=0.05 mm/rot., P=5 MPa, C_{NC}=1% mass.

Fig. 6 – Dependence of roughness on blank rotation speed

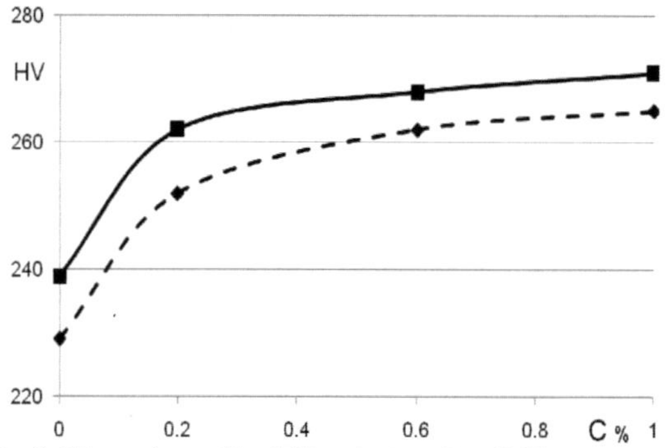

- - S=0.05 mm/rot., V=165 m/min., P=4MPa
—S=0.05 mm/rot., V=165 m/min., P=5 MPa

Fig. 7 – Dependence of hardness on UDAG-S concentration

Roughness reduction of processed surface after increasing of tool advance is happened because the degree of loading on to the processed zone is also reduced. Studies showed that blank rotation speed does not influence much on the surface roughness when hydraulic smoothing (Fig. 9). Pressure elevation of operating liquid provides greater plastic deformation of metal and material hardness and, as a result, hardness enhancement of surface layer.

When studying the hypothesis "about copper film formation", steel samples with copper coating (Fig. 10) no less than 1 mcm thickness (Fig. 11) were obtained after processing.

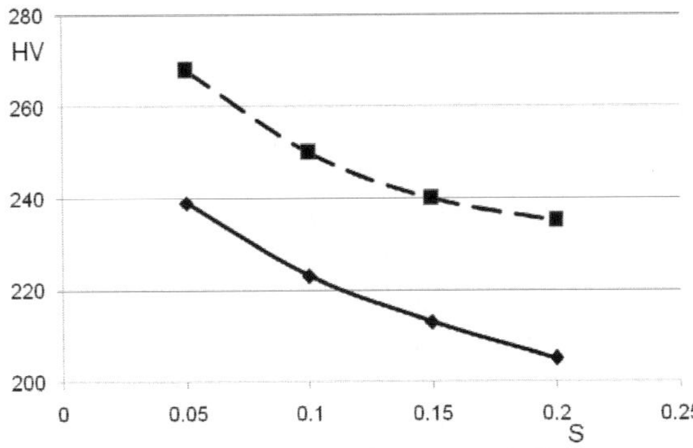

- - -V=165 $m/min.$, P=5 MPa, C_{NC}=0.6% $mass.$
——V=165 $m/min.$, P=5 MPa, C_{NC}=0% $mass.$

Fig. 8 – Dependence of roughness on tool advance

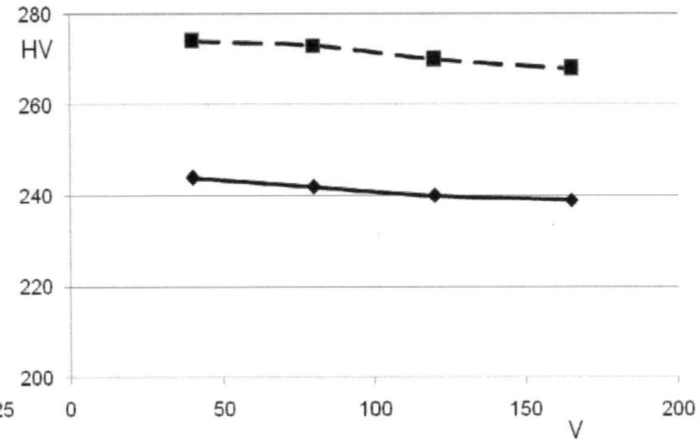

- -S=0.05 mm/rot, P=5 MPa, C_{NC}=0.6% $mass.$
—S=0.05 mm/rot, P=5 MPa, C_{NC}=0% $mass.$

Fig. 9 – Dependence of roughness on blank rotation speed

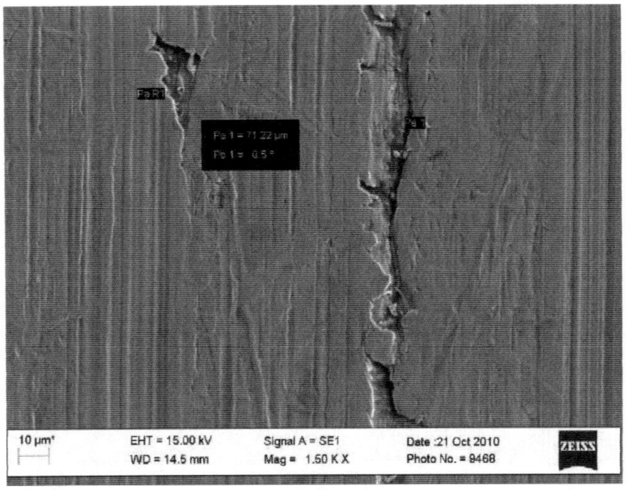

a)

b)

Fig. 10 – Surface of smoothed steel samples a) without Cu; b) with Cu

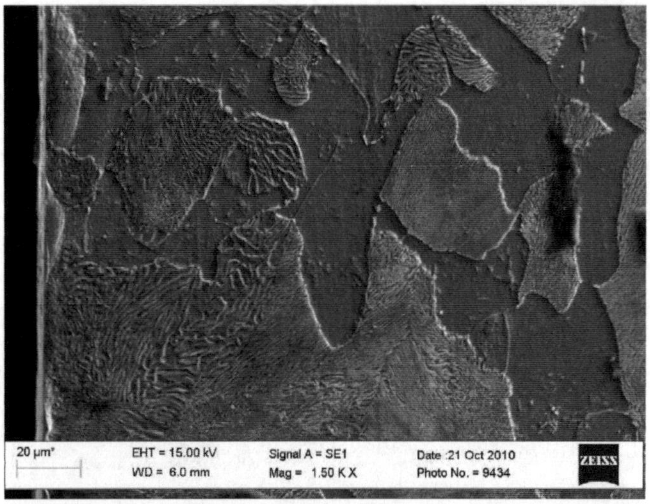

a)

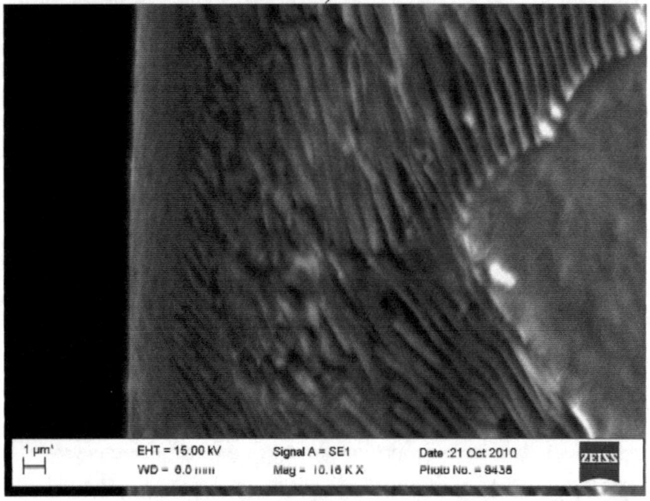

b)

Fig. 11 – Structure of steel sample with copper coating: a) ×1500; b) ×10160

4. Conclusion

The results after conducted studies show that process liquid parameters and their control influence much on the formation of surface layer of the processed articles. During theoretical and pilot studies of hydraulic smoothing, the following conclusions were made:

– methodology elements of technological quality assurance of articles when hydraulic smoothing are presented;

– after experimental demonstration of the supposed hypotheses, the possibility to improve quality parameters of the surface by introduction of friction modifiers into the processing zone is shown, as well as the formation of antifriction coatings on the machine parts when hydraulic smoothing;

– influence dependences of processing modes on roughness and microhardness of surface layer is defined, including the introduction of diamond-carbon particles (UDAG-S) into process liquid.

References

[1] Garkunov D.N. Tribotechnology. Moscow: “Machine building”, 1985. 424 p.

[2] Brendel H. Wissensspeicher Tribotechnik VEB FACHBUCHVERLAG, LEIPZIG, 1978. 344 p
[3] Gutman E. M. Mechanochemistry metal and corrosion protection. Moscow: "Metallurgy", 1974. 232 p.
[4] Kosteckiy B. I. Surface strength material of friction. Kiev: Technika, 1976. 291 p.
[5] Kragelskiy I.V. Friction wear and greasing. Moscow: "Machine building", 1979. 358 p.
[6] Bushe N.A. Compatibility friction surface. Moscow: "Science", 1981. 126 p

The Identification of Nanoscale Structures According to a Parameters of Acoustic Structuroscopy Method

N V Ababkov [1,a], A N Smirnov [1,2,b], N V Bykova [1,c]

[1] T.F. Gorbachev Kuzbass State Technical University, 28 street Vesennyaya, Kemerovo, 650000,Russian Federation
[2] Institute of Coal SB RAS, Leningrad Pr., Kemerovo, Russian Federation

[a] e-mail: n.ababkov@rambler.ru, [b] e-mail: galvas.kem@gmail.com,
[c] e-mail: natalie.vrb91@mail.ru

Abstract. The fracture surface of a destroyed steam turbine rotor is studied by acoustic structuroscopy method. The structural-phase state of the metal of the destroyed rotor of a steam turbine is studied using the methods of electron microscopy. It was established that in the areas of control, where the values of the acoustic characteristics have significant differences from the rest of the metal, detected nanocrystalline structure. The possibility of determining the structure of the nanoscale metal by acoustic structuroscopy is shown.

INTRODUCTION

The destruction of the steam turbine rotor №2 PVS JSC "Zapadno-sibirskiy metallurgicheskiy kombinat" as a result of an accident occurred after 194 000 hours of operation. This destruction occurred in the perpendicular relative to the longitudinal axis of the rotor section, the transition from the diameter of 295 mm (15 steps) to the diameter of 280 mm (16 steps). The source of the break is stress concentrator in the area of the keyway of the rotor stage 15 (Fig. 1). Type of fracture surfaces shows the character of fatigue fracture. The turbine type VKV-22-90 made on Nevsky Machine-Building Plant named after V.I. Lenin in 1963 [1].

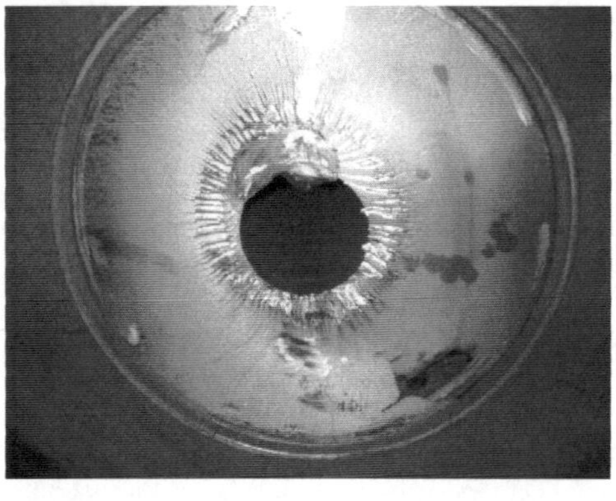

Fig. 1. General view of the fracture surface of the rotor

MATERIAL AND METHODS OF STUDY

The disc for research of the destroyed rotor was cut (sample) parallel to the fracture surface thickness ~ 50 mm. Places cutting thin sections of the disc is determined by acoustic methods, including spectral-acoustic (Rayleigh waves by scanning the surface of the opposite fracture), and an end disk longitudinal waves along the entire circumference [1].

The surface of the sample for research by spectral-acoustic method (CPI "Astron") [2, 3] is divided into zones as rectangles (Fig. 2 a).

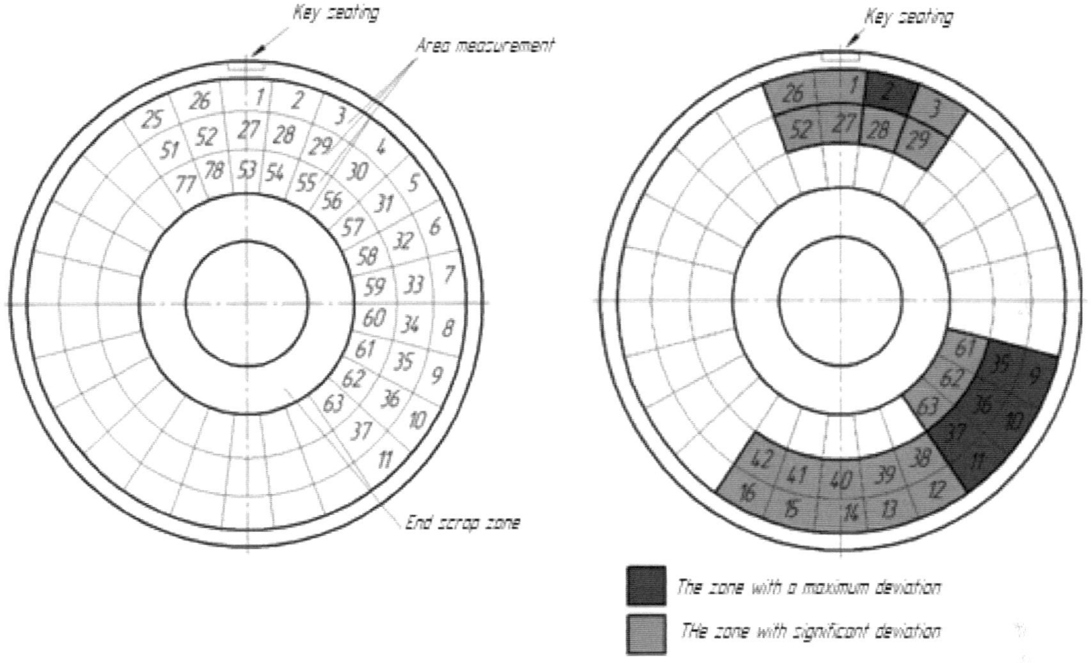

a) б)

Fig. 2. The scheme and measurement results of the acoustic and magnetic characteristics of the surface of fracture the investigated sample:
a - measurement scheme; b - the results of measurements

RESULTS AND THEIR DISCUSSIONS

After carrying out measurement the following picture of the distribution of acoustic characteristics relative to the cross section of the rotor has received. The highest values of the deviations of the measured characteristics detected in the areas of scan 2, 9-11 and 35-37. Also, significant deviations from the average value found in zones 1, 3, 12-16, 26-29, 38-42, 52 and 61-63 (Fig. 2b) [4, 5].

Further, four sites have been chosen with the acoustic characteristics significantly different from the rest of the metal disc (sample). Metalphysical researches conducted in five different areas of the sample on the surface of the opposite fracture and at a distance of 0.5 mm from the fracture. Schematically, the position of these areas relative to the fracture surface is shown in Fig. 3 [1].

It was found that steam turbine rotors made of steel 40H2N2MA (GOST 4543-71). The analysis is performed on optical-emission spectrometer «Q4 Tasman» [1].

X-ray diffraction analysis (XDA), diffraction and transmission electron microscopy on thin foils (TEM) is used in the paper. Research by using the X-ray analysis was performed at room temperature on a DRON-7 [6]. Electron microscopic studies were carried out on a transmission electron microscope EM-125 (without the goniometer). The electron microscope was set to a high resolution, at an accelerating voltage of 125 kV. Working increase in the column of the microscope was × 25.000. Determining the sizes and volume fraction of the phases present in the material carried by the images confirmed by microdiffraction paintings and dark-field images obtained in the reflexes of the respective phases [7].

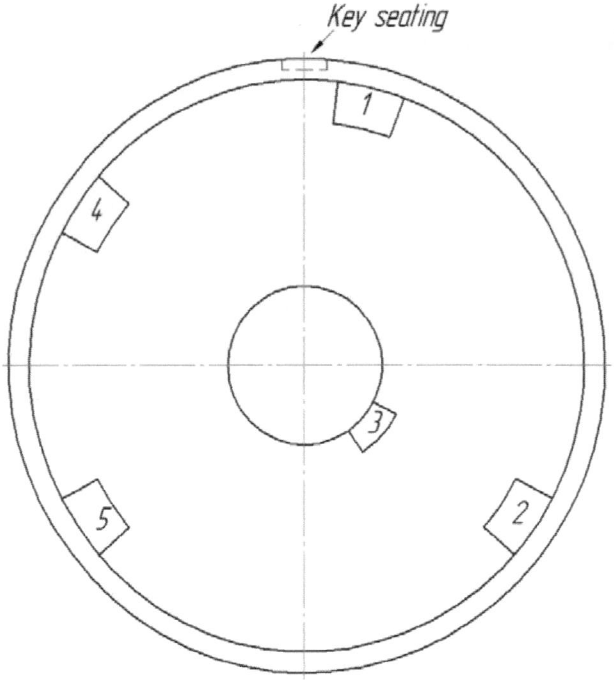

Fig. 3. The scheme places cuttings samples for research

As studies have shown, the basis of all of the samples is α-phase. Its volume fraction at any point of the product is at least 95%. α-phase - a solid solution of both the interstitial elements (C, N etc.) and substitution elements (Mn, Cr, Si, etc.) in α-Fe with a BCC crystal lattice. The basic units take, first of all, the iron atoms, as well as substitution elements, the internodes (mainly tetrahedral internodes) is partially occupied by carbon atoms. X-ray diffraction pattern obtained from different samples are identical.

Crystal lattice parameters α-phase at all points in samples studied are shown in Table. The same table shows the phase composition, microdistortions and sizes coherent scattering areas (CSA), obtained by X-ray diffraction analysis (XDA). As can be seen from Table, at any point in the product crystal lattice parameter is 0.2872±0.0002 nm, ie, slightly higher than the lattice parameter of pure α-Fe. The crystal lattice parameter α-Fe is 0.2866 nm. This means that although α-phase and consists essentially of α-Fe, but in the solid solution except for the substitution elements (Mn, Cr, Ni, Mo, Si) and is also an interstitial element - carbon.

Electron microscopic researches have shown that regardless of research place the main phase component (matrix) investigated steel is α-phase. Morphologically α-phase in different samples are generally present in the form of: 1) a bainite (least defect of the material); 2) a fragmented α-phase with anisotropic fragments; 3) fragmented α-isotropic phase with fragments and 4) a nanocrystalline structure.

Table

The data X-ray diffraction analysis

Sample number		The phase composition	Crystal lattice parameters α-Fe, *a*, nm	Microdistortions, Δd/d	Size CSA P, nm
№ 3		α-Fe	0.2870_5	0.0005	50
№ 1	№ 1-1	α-Fe	0.2870_0	0.0022	>100
	№ 1-2	α-Fe	0.2870_7	0.0014	56
№ 2	№ 2-1	α-Fe	0.2872_0	0.0018	>100
	№ 2-2	α-Fe	0.2871_0	0.0010	75
№ 4	№ 4-1	α-Fe	0.2870_0	0.0020	>100
	№ 4-2	α-Fe	0.2870_0	0.0010	92.5
№ 5	№ 5-1	α-Fe	0.2873_4	0.0010	90
	№ 5-2	α-Fe	0.287_5	0.0005	50

The nanocrystalline structure is not present in all the investigated points of the product. Microdiffraction pattern is a ring. Example electron microscopic image and microdiffraction pattern shown in Fig. 4.

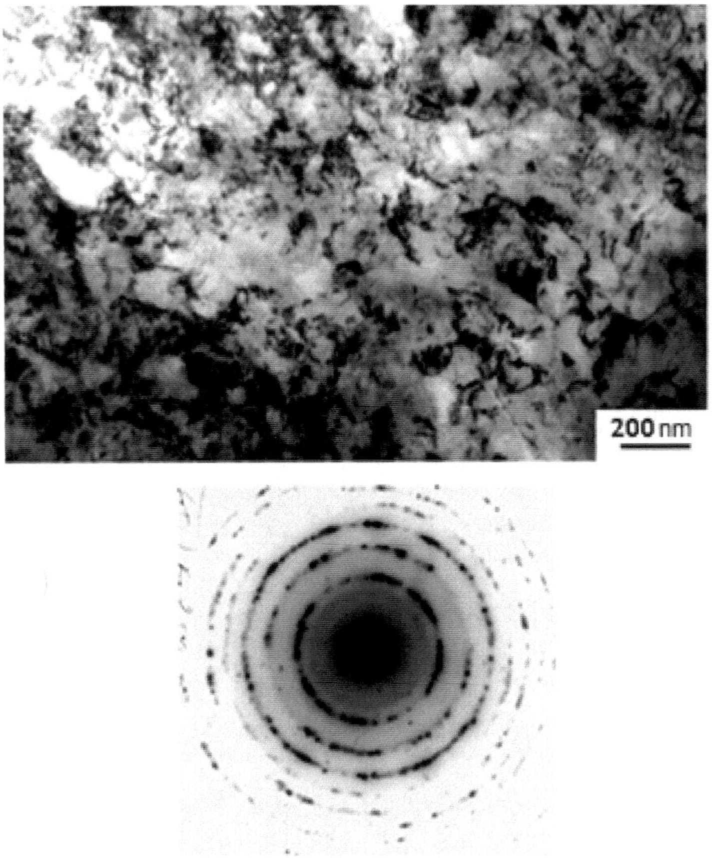

Fig. 4. Nanocrystalline structure. Electron-microscopic image

CONCLUSION

Acoustic structurescopy - it's the definition of composition, grain size, hardness, anisotropy properties of impurities in the material by using of acoustic methods of nondestructive testing.

Electron microscopic study were subject only those areas in which the performance of acoustic characteristics significantly different from the rest of the metal disc. It follows that the parameters of acoustic structuroscopy can judge about the presence of nanosized structure metal.

ACKNOWLEDGEMENT

The work is financially supported by project RSF No.14-19-00724.

REFERENCES

[1] Aleshin, N. P. Fizicheskie metody nerazrushajushhego kontrolja svarnyh soedinenij. – M. : Mashinostroenie, 2006. – 366 p.

[2] Mitenkov, F.M. O novom metode kontrolja povrezhdaemosti materiala oborudovanija JAJEU i apparatno-programmnyh sredstvah dlja ee realizacii / Mitenkov F.M., Uglov A.L., Pichkov S.N., Popcov V.M. // Problemy mashinovedenija i nadezhnosti mashin, 1998. – №3. – P.3–9.

[3 Ababkov, N. V. Sovremennoe metodicheskoe obespechenie dlja ocenki sostojanija metalla potencial'no opasnogo oborudovanija. CHast' 2. Spektral'no-akusticheskij metod kontrolja / N. V. Ababkov, A. V. Benediktov, A. N. Smirnov i dr. // Vestn. KuzGTU. 2010. – № 5. – P. 101–106.

[4] Dolzhanskij, P. R. Kontrol' nadezhnosti metalla objektov kotlonadzora: sprav. posobie. – M.: Nedra, 1985. – 263 p.

[5] Smirnov, A. N. Mikrostruktura, polja vnutrennih naprjazhenij i akusticheskie harakteristiki metalla razrushennogo rotora parovoj turbiny / A. N. Smirnov, N. V. Ababkov, JE. V. Kozlov i dr. // Izvestija vysshih uchebnyh zavedenij. Chernaja metallurgija. – 2014. – V. 57. № 10. – P. 67–71.

[6] Kontrol' metalla v jenergetike / pod. red. V. G. Borisova [i dr.]. – Kiev : Tehnika, 1980. – 134 p.

[7] Ivanov, JU. F. Gradientnye strukturno-fazovye sostojanija, formirujushhiesja v stali 08H18N10T pri mnogociklovoj ustalosti do razrushenija / JU.F. Ivanov, V.E. Gromov, S.V. Gorbunov i dr. // Fizika metallov i metallovedenie. – 2011. – V. 112. – № 1. – P. 85–93.

Forecasting of the Holes Shape Accuracy of Thin-Walled Body Parts Through the Application of Artificial Neural Networks

V A Kalistru, A G Ovcharenko, A M Firsov

Biysk technological Institute (branch) Altai state technical University n.a. I.I. Polzunov, Biysk, Trofimova street, 27

e-mail: mrsi@bti.secna.ru

Abstract. In this paper is proposed a method to predict deviations of accuracy of a basic shape of holes due to a thermal deformation (TD) at the design stage of the technological process. Proposed method of control of technological process consists of two stages: the first stage (auxiliary) is based on the finite element method (FEM), the second (main) - on the modeling of artificial neural network (Ann). In this paper is developed an algorithm of calculation of input and output parameters of the network using LS-DYNA. A structure of the Ann to predict and to adjust a trajectory of a movement of a tool at the preparation stage of the technological process for work pieces included in group process is invented in this article.

1. Introduction

During the competition any business should focus on release of quality products, which must meet the requirements of a customer and a consumer. Application of modern technologies in a field of computer simulation contributes to achieve these goals. The use of artificial neural networks in combination with the method of finite elements will allow to improve indicators such as the manufacturing accuracy and surface quality.

In modern conditions of production mechanical treatment by cutting still does not meet the requirements, resulting in an inclusion in the process of costly finishing operations. Therefore, improving of blade processing through the use of information technologies, which allow to increase performance and to provide high quality of products, is an urgent task.

As objects of study were selected base holes of thin-walled body parts, because they imposed high demands on dimensional accuracy (6-8 degree), deviation (4-5 grade), surface roughness (RA=1.6 to 0.08 μm) (figure 1) [1].

Thin-walled housing parts in order to facilitate the mass are manufactured by casting from aluminum alloys. Through cooling of castings due to a different thickness of walls is forming a heterogeneous structure, non-uniform hardness, which results in a formation of unstable mechanical properties of a surface, different roughness and a deviation of a size and a shape of a surface during processing. Accuracy of the shape is one of the important requirements in the treatment of underlying holes of body parts, because they imposed increased demands for durability.

Significantly on accuracy of processing also affects a non-uniform hardness, which leads to a change of cutting forces and different elastic deformations of elements of technological system. This results in a scattering of size and a deviation of the shape of the hole.

IOP Conf. Series: Materials Science and Engineering **126** (2016) 012007 doi:10.1088/1757-899X/126/1/012007

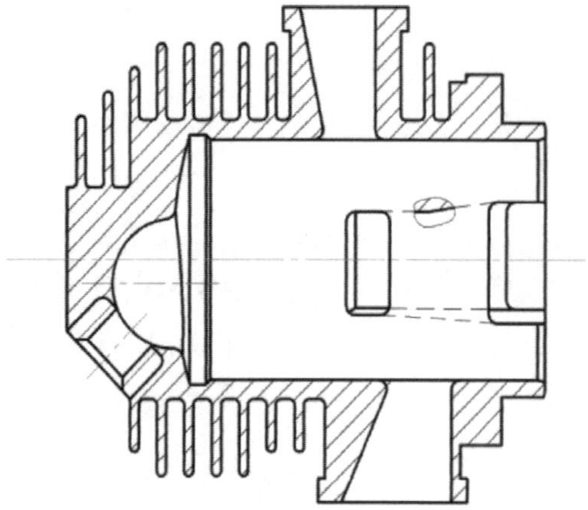

Figure 1. The thin-walled basic part.

During the measurement of hardness the Converter type USDT-100N was used. The surface of the cylinder is divided into 6 equal parts at intervals of 11mm. Over the length of a border of each piece in different places were made 5 measurements, average values were recorded in table (table 1).

Table 1. The study of hardness.

The number of the workpiece	Hardness, MPa					
	1	2	3	4	5	6
1.1	327	302	361	364	322	320
1.2	289	337	313	373	410	405
2.1	299	329	305	364	337	315
2.2	272	296	317	299	337	342

For hardness values of workpieces mean values were calculated, which resulted in the graphs (figure 2).

Variable hardness leads to a change of cutting forces and different elastic deformations of elements of technological system. This results in a scattering of size and a deviation of the shape of the hole. It can be assumed that variable hardness occurs during a crystallization of a metal due to varying wall thickness of the workpiece and it is one of the negative factors that affect the expected accuracy.

The expected machining accuracy is a cumulative result of many elementary inaccuracies. Each of them contributes its share to the total inaccuracy and at the same time indirectly or directly affects the other elementary inaccuracies [2].

Share of elementary inaccuracies in the total inaccuracy for the most common cases is approximately [2]:

• setting inaccuracy during the processing of holes of these body parts: 2...40%;
• inaccuracies due to elastic deformations of the technological system: 0...80%;
• inaccuracies from the settings and tool wear: 5.. .40%;
• inaccuracies due to thermal deformations of elements of TS: 15...70%;
• form inaccuracies of the machined surface due to the geometric inaccuracies of the machine: 1...15%.

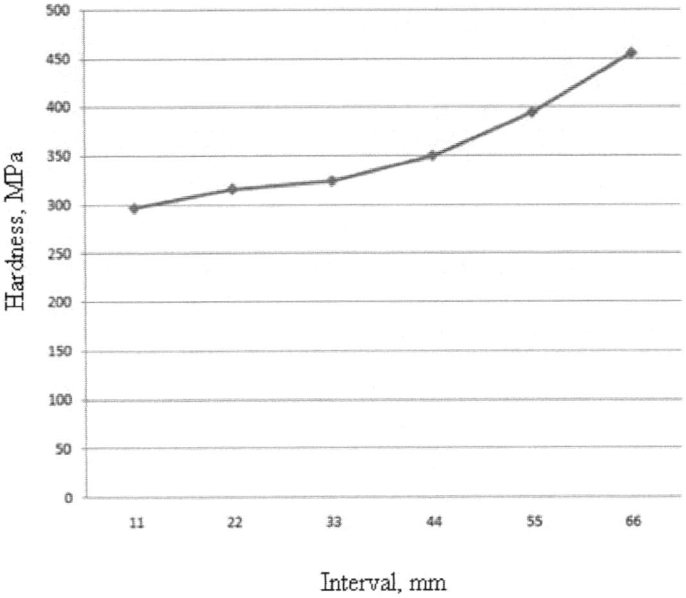

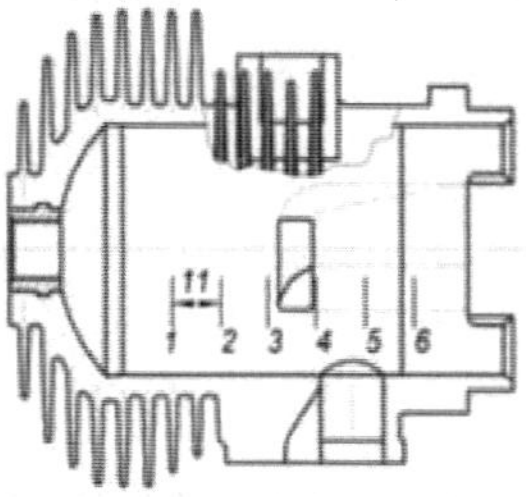

Figure 2. The study of hardness.

For thin-walled parts with variable wall thickness errors, involved with thermal deformation, can account for up to 70% [3]. Consequently a reduction of heat generation during cutting and, accordingly, a decrement in TD of a workpiece are possible mainly by reduction of processing modes, that leads to reduction of performance. Thus, it is necessary to predict the magnitude of thermal deformation at the design stage of a technological process using an integrated approach, that consists of two stages: the first stage (auxiliary) based on the method of finite element analysis (LS-DYNA), the second (prediction) stage based on a learning and building of neural networks (Statistica Neural Networks).

2. Theory

Analysis in LS-DYNA simplifies and automates only a calculation of required parameters, but has several disadvantages: greatly increases a duration of preparation of a technological process, even on modern computers calculations are carried out a considerable time; complex preparatory phase; it is required a high qualification of specialists. Thus, the phase of calculation using the finite element method is complement and necessary for a calculation of input and output parameters of the artificial neural network used to predict deviations of a precision of a hole shape. For this purpose one detail is selected on which simplified models are built (training set), that include the following specifications: hole diameter (D), long hole (L), difference between minimum and maximum wall thickness (a), a workpiece material, a cutting tool material. To improve a performance of calculation, instead of the cutting tool cutting wedge is used, because TD is calculated only for harvesting (figure 3).

One of the most important stages of finite element analysis is building a finite element mesh, in this case accuracy of calculations depends on the correct choice of types and sizes of elements. In LS-DYNA there are two main methods: constructing the arbitrary grid and building the ordered mesh. Practice of calculations using the finite element method demonstrates that the most preferred is the ordered grid; furthermore a rectangular grid is more favorable than a triangular mesh element. A fine mesh is required where it is expected a large gradient of deformations or tension, for example, due to the presence of hubs, cracks or ruptures of cell references. A large mesh can be used in areas with little

IOP Conf. Series: Materials Science and Engineering **126** (2016) 012007 doi:10.1088/1757-899X/126/1/012007

varying relative deformations and tension, as well as in areas not of particular interest for the calculation.

In the study of fluidity any field there are two ways of splitting it into elements –elements are fixed in space and provide for the flow of material through them (the Euler equation), or elements are fixed on the flowing material, so as to move together with it (the Lagrange equation) [4]. On one hand the advantage of analysis by the method of Lagrange is easy to monitor changes of material properties, on the other hand is a computational complexity. Simplicity of computation by the Euler method does not allow implementing the modeling of the chip formation fully in conditions of large displacements of free surfaces.

The element type must conform to the task from the point of view of physics of process and have the appropriate mathematical description and degrees of freedom. So, to perform finite element analysis of cutting process, it is necessary to use several elements and models focused on their task.

To determine the temperature and its distribution in the cutting zone choose the element SOLID70 used for temperature analysis, and which allows calculating of heat transfer and computation the value of the temperature in the nodal points. This element supports heating by the friction, nonlinearity of the material and allows specifying a core set of its properties: density, thermal conductivity, heat capacity, thermal expansion coefficient, etc. As the load is possible to use different heat sources. Calculation results shape the temperature field and heat fluxes.

Figure 3. The finite element method with specified conditions.

Thus the algorithm is constructed as follows:

Auxiliary stage [5-7].

1) Construction of the 3D model and the cutting tool (insert).

2) Generation of a grid

3) Export to LS-DYNA and assignment of material properties and contact conditions.

4) The task of coordinates of the tool path and a frequency of rotation of the workpiece (cutting data).

5) Perform of elastic-plastic analysis (the tool movement and rotation of the workpiece).

6) Calculation of a temperature of the workpiece in a cutting zone. Determining of the significance of thermal strains.

7) Analysis of deformations and deflections shape of basic hole of thin-walled housing part.

8) If the deviation is above acceptable, then back to item 4.

If the deviation is within the valid value, then a file with data is created and exported to a database, and from there is populated the artificial neural network model and training.

The main phase [8].

According to results of the auxiliary stage the artificial neural network is constructed and it consists of 12 independent parameters, which are given per input: submission (s), cutting speed (V), depth (t), the material of the cutting, the workpiece material, hole diameter (D), difference between minimum and maximum wall thickness (a), hole length (L), the ratio of stutter (CRC), hardness of parts, thermal conductivity of the workpiece material, the temperature in the cutting zone (T). Cutting fluid is applied during whole treatment.

The output will have four adjusted parameters: feed rate (dS), cutting depth (dt), cutting speed (dV), allowable thermal distortion (dT) (figure 4).

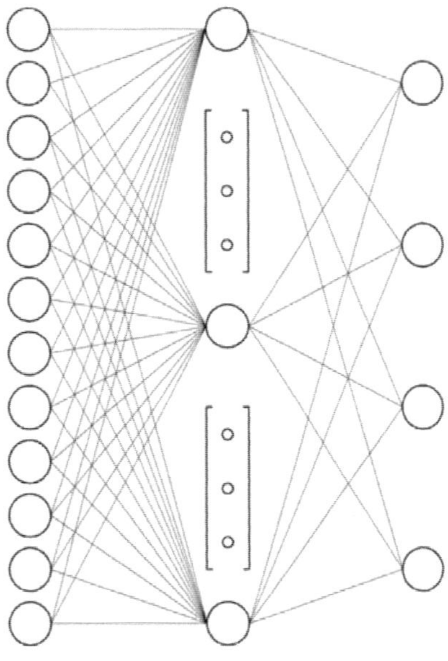

Figure 4. A structure of the artificial neural network for forecasting of deviations of hole shape.

In further the network is tested with input values of input parameters not involved in training sample, and results of the response of the artificial neural network with test values of output data are compares. The actual value of the thermal deformation and the value of the thermal deformation, calculated by using the artificial neural network, are analyzed. If the error is within the acceptable range, a program code of this network prints on one of the programming languages (e.g. C++). It is developed the program, the core of which consists of code of the neural network, and the interface is used to enter calculation parameters and obtain corrected cutting data.

3. Results and discussion

When predicting values of TD the method based on the artificial neural network is able to consider not only a large number of factors and their mutual influence on each other. It, theoretically, has a possibility of approximating of arbitrarily complex dependencies that enables to predict accuracy of machining directly depending on values of input parameters of the process. Disadvantages of this method are: the use of the same type of tool materials and workpiece materials (in our case solid alloys of various grades and aluminum alloys of varying hardness and density). So, forecasting of the technological process of processing of base holes of thin-walled body parts using this neural network will work within boundaries of the group of parts close by design features.

4. Conclusion

The proposed method allows at the design stage of technological process to adjust the trajectory of the cutting tool on the basis of temperature deformations of the workpiece, when there is not stock of technological precision, i.e. there is a possibility of a manufacturing defect due to excess of the error processing over the permissible value. Application of the artificial neural network method will reduce time of technological preparation when designing the group process.

Main factors from the point of view of economic efficiency from implementation of the proposed method in the production are: reduction of time and costs of development of new technology, reduction of costs related technological processes of manufacturing parts and product assembly, increase reliability, durability and serviceability of products. Reduction of terms of development of new technology has important economic value. Long time of design and development of new products reduce efficiency of social production.

References

[1] Tatarkin, E.Yu. Ensuring of the holes accuracy in the processing of small-sized body parts on lathes [Text]. / E.Yu. Tatarkin, A.M. Firsov // Polzunovskiy vestnik, 1/1-2012. – P. 286-290
[2] Kolev, K.S. The accuracy issues when cutting metals [Text]. /K.S. Kolev – M. Mashgis, 1961. – 134 p.p.
[3] Kalistru, V.A. The problem of technological ensuring the accuracy of shape of holes of thin-walled body parts / V.A. Kalistru, A.M. Firsov//Actual problems in machine building: proceedings of the second international scientific and practical conference – Novosibirsk, 2015.– P. 92-95.
[4] Morozov, E. M. ANSYS in engineer's hands. Fracture mechanics. / E.M. Morozov, A. Yu. Muizemnek, A.S. Shadskiy // Publisher: Lenand, 2014. – 456 p.p.
[5] Maekawa, K. and Maeda M. Simulation analysis of three-dimensional continuous chip formation processes – FEM formulation and a few results. – Japan Soc. Prec. Eng., 1993. 59(11), 1827-1833.
[6] Zienkiewicz, O. C. The Finite Element Method in Engineering Science 2nd edn, Ch 18. London : McGraw-Hill, 1971 – 1988.
[7] ANSYS 12 Theory Reference, LS-DYNA User's Guide / SAS IP, Inc.,2009.
[8] Haykin, S Neural Networks A Comprehensive Foundation Second Edition, McMaster University Hamilton, Ontario, Canada, 1999. – 823 p.p.

Method of refusals for technological preparation of group processing of basic holes

E Y Tatarkin, A M Firsov, A G Ovcharenko, V A Kalistru

Biysk Technological Institute, Trofimova str. 27, Biysk, 659305, Russia

e-mail: mrsi@bti.secna.ru

Abstract. Analysis of the group stages of the preparation process for the body parts has shown that the existing methods of design does not account a number of factors that give rise to unacceptable deviations (potential failure) the output process parameters. Algorithm design of group process with the use of methods of "Analysis of the types and consequences of failure" and design search is proposed. This allows the design phase to identify potential failures, to develop measures to prevent such failures and to provide the required output parameters.

Introduction

The efficiency of modern technological equipment is determined by the degree of its utilization. Increased equipment utilization is possible through the use of batch processing components, the principles of which were developed by S.P. Mitrofanov, A.P. Sokolovsky [1, 2, 3] and found in the works of B.M. Bazrov [4, 5].

Today, the group handling is not widely used in the manufacture of case details due to the large variety of structural forms, materials and methods of production of billets. It is not possible to create complex parts for which is possible to use standard technology solutions. One solution to this problem is the association of case details in groups of design and technological features basic holes, the complexity of processing that can be more than 60% of the total labor input.

A distinctive design feature of the base hole is not always possible to apply technological solutions for the processing of basic holes of all details within the group. This is due to the fact that structural and other features not included in the normative and technical documentation. Therefore processing of such details, even over a single technological process can lead to unacceptable deviations (potential failure) [6, 7]. In this case, the model adopted technological solutions specific to individual processes do not provide the required output parameters of group process step in the manufacture of all details of the group. When processing some details of the group there are potential failures, such as failure to comply with a geometric precision machined surfaces, performance, costs, etc.

Troubleshooting failure by the trial moves and measurements during debugging, or when the process significantly increases the complexity. Therefore, it is necessary to predict the appearance of potential failures at the design stage of technological processes (TP) for all the details combined into a group and to take measures to prevent them.

Materials and methods

Prediction of output parameters TP and their potential failures may be performed using analytical, statistical and expert methods of estimation. However, in multiproduct production conditions when designing batch processing application of these methods is not always possible, because a group can be combined by various design details.

An analytical method may be time-consuming and not always adequate for all the constructions of details, the results of statistical studies are often lacking, and the peer review only allows a qualitative assessment of the output parameters. Therefore, you need a method to identify potential failures arising in the performance of the process. This will identify the main causes of failure and their possible consequences, to develop action to eliminate these causes or mitigate the consequences.

The authors proposed a perspective method of "Analysis of failure modes and effects", which has been used successfully to assess the reliability of various facilities and technology [8-10].

However, the use of this method in the design of group processes (GTP) is unknown and requiring additional research.

According to [11] the design of group process involves a number of steps that can be grouped into three basic 1-analysis of the original data and the creation of a grouping of details of complex details; 2 - design of TP for complex details and assessment of its results; 3 - provision of conditions to TP and documentation. However, the design of GTP handling of details with distinctive design features necessary to introduce additional steps.

These stages are connected with the analysis of the influence of special structural components parts within the group, assessing their impact on the output parameters of the TP.

Changing the conditions of the TP, designing technological and technical solutions taking into account the structural features of group details provide required output parameters for all the details.

The implementation of these proposed steps are made using the methods of "Analysis of the types and consequences of failure" and design search.

Results and discussions

Algorithm of design of GTP using the above methods is presented in the form of a scheme consisting of 7 stages (Figure 1). In the first stage the original data for the design were analyzed: drawings of parts with specifications for dimensional accuracy, the deviation of the surface roughness, etc .; software task and deadline for the task; financing offer and the cost of technology.

Grouping of parts on different grounds are produced after the analysis of the original data and the award of the contract for the manufacture of parts. When you combine a group of different design details it is necessary to identify similar in design and technological features surface and create an integrated surface that combines all the features of surfaces of parts.

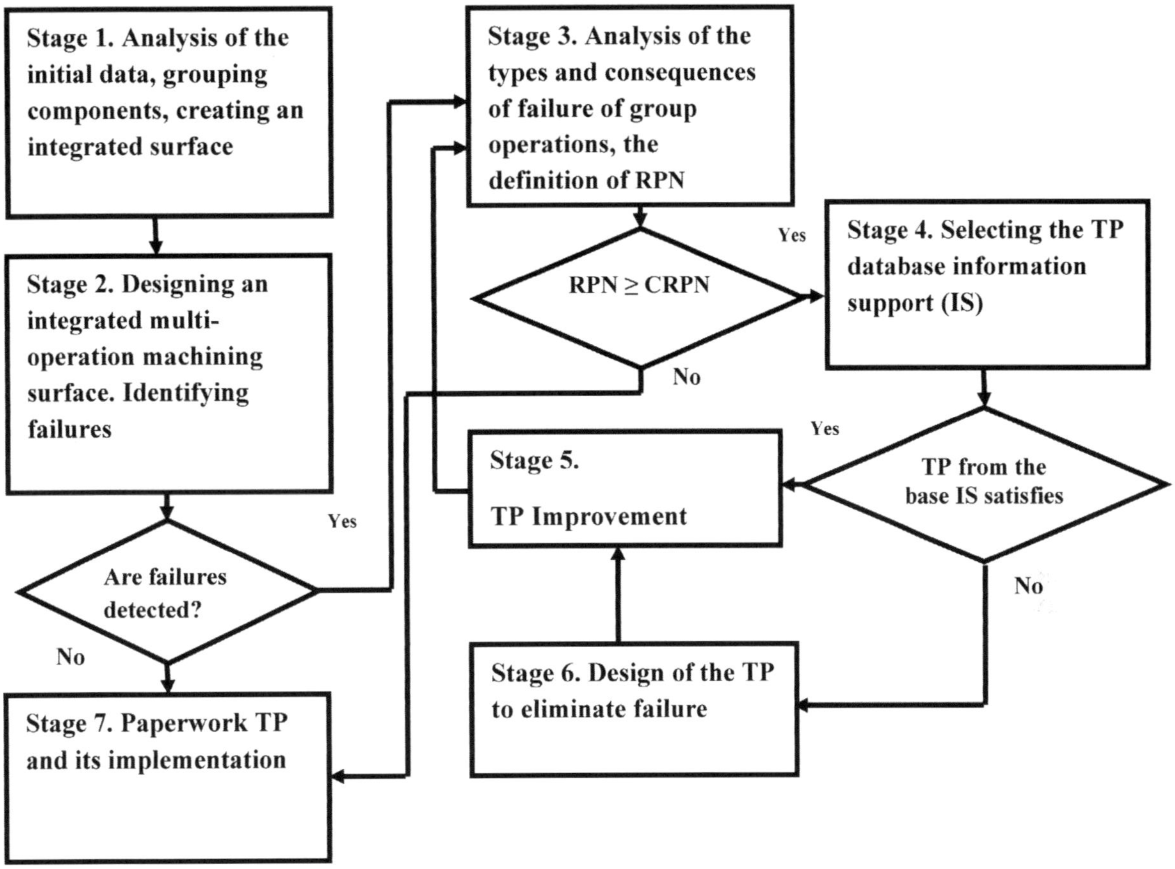

Figure 1. Scheme of the algorithm design of group process step using the methods of "Analysis of the types and consequences of failure" and design search

In stage 2, the group operation and plan when processing a particular surface of the range of parts group are developed. If the group together different parts of the structure, the group operation is developed for integrated surfaces. For them, technology is determined by the sequence of actions to provide the desired process requirements. Thus, a design and technology unit, which consists of integrated surface and process for the group details are created.

Transition plan surface treatment parts, tools and processing modes are selected from a database of known standard solutions, regulated legal and technical documentation. Since the group merged with the details of design features, which are not considered legal and technical documentation, it is necessary to analyze the impact of these features on the process. If there are structural elements or other factors which may lead to potential failures, it is necessary to analyze their effect on the output parameters TP.

The third stage of the design is carried out according to the requirements of the method "Analysis of the types and consequences of failure" and includes a number of actions:

- to establish an expert group to identify potential for failure of the process and their implications;
- the level of importance of the consequences of refusal is determined by 10-point scale (***C***);
- the level of importance of the reasons for the emergence is determined by 10-point scale (***R***);
- the level of significance of detection of failure in the manufacturing process is determined by 10-point scale (***D***);

- for each failure risk priority number equal $RPN = C \cdot R \cdot D$ is calculated. *RPN* has a value from 1 to 1000, a critical risk priority number *CRPN* ranges from 100 to 125;

- number of each priority risk is analyzed. If $RPN < CRPN$, the failure is considered to be insignificant, the design process ends and the paperwork is done (Figure 1, stage 7). If $RPN \geq CRPN$, the failures are considered to be significant and for operations such new technological solutions are developed.

On the fourth stage it is necessary to analyze the availability of technological solutions in the data base of information to support the design of group processes. This database contains information on structural elements of detail that can be the cause of failure, failure modes, and technological solutions that prevent the occurrence of failure wholly or reduce its exposure to acceptable requirements.

This information support is formed from data obtained in the design of TP with similar structural and technological features. If there is a technological solution that was previously developed, it is necessary to carry out its assessment of the criteria, which include a set of requirements in terms of functionality, technology, economy and safety.

If an existing solution is satisfied the requirements, it is used to improve the process (Figure 1, stage 5), which is to change the type of processing, structure, tools equipment, processing conditions, etc.

If the technological solution does not satisfy the requirements, go to stage 6 "Design of technological solutions to address potential failures." In this case, new technological solutions includes the formulation of the problem in the design, development, technical specifications and the design itself. At this stage, methods of search engine design, such as "brainstorming", "heuristic methods", "morphological analysis and synthesis" and others are used. Founded technological solution is evaluated using criteria.

In meeting the required specification technological solution is used to improve the TP (Figure 1, stage 5) and is entered into a database for further information to support the application.

After the improvement of the TP priority the number of risk *RPN* is determined . In the case of providing the desired output parameters for TP it is taken to implement, otherwise the cycle repeats as long as the condition is met $RPN < CRPN$.

Conclusions

An analysis of the stages of the preparation technological process for the group of case details found that the current methodology does not take into account a number of design factors, or other structural nature, which give rise to unacceptable deviations of the output parameters of the process.

An algorithm for the design of group process with the use of methods of "Analysis of the types and consequences of failure" and design search is offered. This allows the design phase to identify potential failures, to develop measures to help prevent the emergence of failures and provide the required output parameters in the implementation of the group process.

The proposed method allows to detect and eliminate potential failures that may occur as a result of the lack of data in the normative and technical documentation on the design of technological operations, the design features of the workpiece and other factors at the stage of technological preparation of production.

References

[1] Mitrofanov S.P. Nauchnaya organizaciya mashinostroitelnyh proizvodstv [The scientific organization of engineering industries]. Leningrad, Mechanical Engineering, 1976. 710 p.

[2] Mitrofanov S.P. Grupovaya tekhnologiya mashinostroitelnogo proizvodstva.Tom 1. Organizaciya grupovogo proizvodstva [Group technology engineering production. V.1. The organization of group production]. Leningrad, Mechanical Engineering, 1983. 407 p.

[3] Sokolowski A.P. Kurs tekhnologii mashinostroeniya. Chast 1. Obshie voprosy tekhnologii mehanicheskoi obrabotki [Course of mechanical engineering. Part 1. General questions machining technology]. Moscow- Leningrad, Mashgiz, 1947. 435 p.
[4] Bazrov B.M. Modylnaya tekhnologiya v mashinostroenii [Modular technology in mechanical engineering]. Moscow, Mechanical Engineering, 2001. 368 p.
[5] Bazrov B.M. Tipovaya tekhnologiya v sovremenyh usloviyah [Types of technology in modern conditions]. Naukoemkie tekhnologii v mashinostroenii - High technologies in mechanical engineering, 2014, no. 4, pp. 44-46.
[6] GOST 27.202 - 83 Reliable technology. Technological systems. Methods for assessing the accuracy of the parameters of quality of manufactured products. Moscow, Publishing Standards, 1983. 350 p.
[7] GOST 27.310 - 95 Reliable technology. Analysis of the types, effects and criticality of failures. The main provisions. Moscow, Publisher standards, 1996. 20 p.
[8] GOST R 51814.2-2001 Quality systems in the automotive industry. The method of analysis of species and the effects of potential defects. Moscow, Publishing standards, 2001.- 19 p.
[9] Firsov A.M., Vdovin A.V., Timokhovitch I.V. Proektirovanie tekhnologicheskogo procesa obrabotki rezaniem na osnove ucheta otkazov tekhnologicheskoi sistemy [Design process machining on the basis of accounting failures of technological systems]. Mekhaniki XXI veku - Mechanics of XXI century, 2013, №12, pp. 155-159.
[10] Tatarkin E.Y., Firsov A.M., Kalistru V.A. Obespechenie tochnosti bazovyh otverstiy korpusnyh detaley pri grupovoi obrabotke putem prognozirovaniya parametricheskih otkazov [Ensuring the accuracy of basic holes of body parts in the group treated by predicting parametric failures]. Naukoemkie tekhnologii v mashinostroenii - High technologies in mechanical engineering, 2015, №5, pp.15-29.
[11] R 50-54-93-88 Recommendations. Development and application Processes. Moscow, Publishing standards, 1988. 15 p.

IOP Conf. Series: Materials Science and Engineering **126** (2016) 012009 doi:10.1088/1757-899X/126/1/012009

Generalized Mathematical Model Predicting the Mechanical Processing Topography

S L Leonov [1, a], A M Markov [2, b], A B Belov [3, c], N Sczygol [4, d]

[1,2,3] Polzunov Altai State Technical University, 46, Lenina prospect, Barnaul, 656038, Russia
[4] Czestochowa University of Technology, ArmiiKrajowej 21, 42-201 Czestochowa, Poland

[a] sergey_and_nady@mail.ru, [b] andmarkov@inbox.ru,
[c] alexbelyj1209@yandex.ru, [d] norbert.sczygiol@icis.pcz

Abstract. We propose a unified approach for the construction of mathematical models for the formation of surface topography and calculation of its roughness parameters for different methods of machining processes. The approach is based on a process of geometric copy tool in the material which superimposes plastico-elastic deformation, oscillatory occurrences in processing and random components of the profile. The unified approach makes it possible to reduce the time forcreation of simulated stochastic model for a specific type of processing and guarantee the accuracy of geometric parameters calculation of the surface. We make an application example of generalized model for calculation of roughness density distribution Ra in external sharpening.

Introduction

The surface tool topography has a significant impact on its serviceability and life cycle [1 - 4]. Detail topography allows considering geometric parameters of its surface with different detail levels. It includes the size, form deviations, waviness and roughness of the surface. These indicators and their parameters have a significant effect on the detail serviceability. Having the surface topographyit is not difficult to get any parameters of geometric ones. Therefore, when designing the model of the detail, irrespective of the type of processing, it is necessary to focus on its topography – describing the geometry of the surface as a whole.

The measurement process is very laborious, despite there is special equipment for measuring the topography, as well as detail profilograms. Therefore, it is more efficient to predict the topography by means of mathematical modeling for the process of its formation. Due to empirical models are of limited use, and different conditions of machining process have big differences, it is relevant to create a common approach predicting the topography of the finish surface, regardless of the processing type [5].

Theory.

The main events, identifying the topography of the finish surface after the mechanical process, are [1, 6]:

1. geometric coping of the working part of the tool in the material surface of the workpiece;
2. vibrational movement of the tool relative to the workpiece;

3. plastico-elastic deformation of the surface layer of the processed material;
4. random digging of the processed material.

These phenomena are also accompanied by surface damage [7], which affects both size and form of its working parts and the dynamics of the cutting process [8] (cutting force, temperature, etc.). Account must be taken on the sticking of material workpiece surface to the tool (built-up edge) may also change its geometry and dynamics of cutting. Furthermore, the geometry of the tool, the cutting mode settings, hardware settings and devices, physical and mechanical properties of the worked and tool material are not constant and contain both dynamic and random components [5].

In accordance with it, the simulated model of the topography forming machining operation is based on the *geometrical copy process* of tool profile to the product material. From the theory of sets point of view, the cutting process can be viewed as a process of removing multiple sets of M_{APTj} points, given by the actual profile of the tool, made of workpiece material (set M_i). Summary actual profile of the tool may consist of several subsets (e.g. the cutting portion profile of the grinding wheel is determined by the profiles of its constituent grains, and the cutting part of the cutter profile – its teeth). Therefore, union of subsets is used to describe. At every point in timei in the original profile (a set of M_{i-1}) cut the intersection of this set with a set of the tool actual profile. This process is describedin the formation of a new profile M_i:

$$M_i = M_{i-1} - \left[\left(\bigcup_j M_{APTj}\right) \cap M_{i-1}\right] \qquad (1)$$

In the process of modeling M_{APTj} plurality gradually changes due to, for example, tool wear. The number of elements of this set depends on the mode and the kinematics of the cutting. In the simulation processing blade tool (turning, milling, etc.) set $M_{ФПИj}$is given in the cutting plane. Profile parameters have stochastic components, and the density of their distribution depends on the accuracy of the instrument and its home base in the device. When simulating abrasion stochastic nature of the process inherent cut: grain instrument have a random distribution of its depth and surface, grains have a random size and shape. But even with the blade, and abrasion processing in the formation of the profile parts there involved copying process of geometric cutting elements (incisor teeth milling, grains of the abrasive tool).

The process of the tool profile copying is accompanied by a *process of elastic deformation as* processed material, and the instrument. High temperatures in the contact zone of the tool with the workpiece (especially when grinding) lead to smoothing of the resulting profile. When the blade processing the speed factor and the temperature is lower, but high specific cutting force deforms come offroughness ridge by changing its shape. In the actual cutting processthere are two factors: smoothing of the profile, and nonlinear deformation. Fundamental process depends on the specific cutting conditions.

Influence of vibrations during machining is not completely clear. The sources of vibrations are as fluctuation of technological system elements from drives and unbalanced masses (e.g. an imbalance of the grinding wheel or parts), and the cutting process – self-oscillations (regenerative, coordinate communication, nonlinearity, etc.). In the latter case, the analysis requires the study of self-oscillations in the feedback system. Therefore, the output of the model are, in addition topography, and the dynamic characteristics of operation: cutting force, the speed and acceleration of movement of elements of the technological system, etc.

The processes of profile forming and the surface topography are *random components* related to pickups andpits of the work material. There are instances when the process causes the substantial surface roughness. Therefore, the development of stochastic models supposes taking into consideration the random component of the surface topography.

Fig. 1 shows the structure of a generalized simulation stochastic model of the topography formation. The structure takes into account:

1. The topography formation of the processed surface parameters affects the tool, equipment, devices and cutting conditions.

2. Stochastic parameters of the subsystems are additive.
3. The parameters of the subsystems are defined as dynamic properties of the simulation model (feedback).
4. The surface topography is formed by four phenomena: the geometrical copy, the impact of vibration, plastico-elastic deformation and proper random component.
5. Any geometric parameters of the surface layer are formed by the output function of the surface topography.

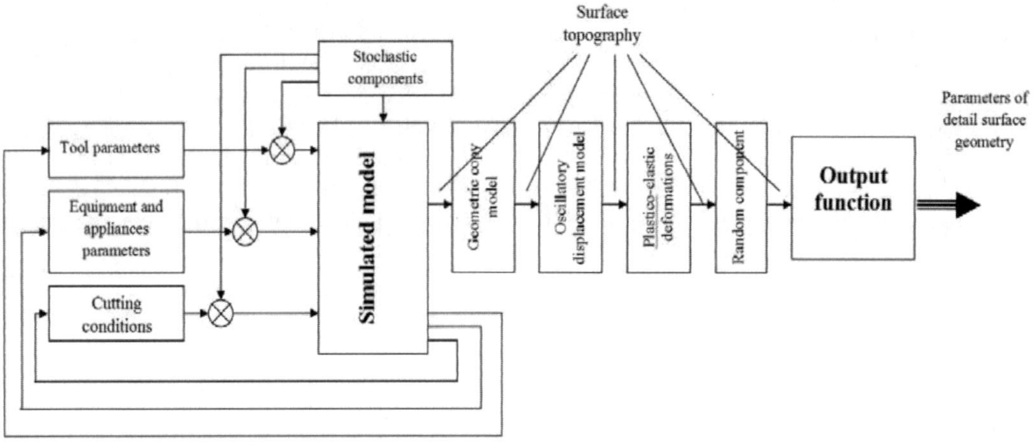

Figure 1 - The structure of generalized stochastic simulation model of mechanical processing

According to developed structure we have implemented stochastic simulated models of a number of mechanical processing. For blade machining they are represented by lathe turning and face milling. For abrasion – by circular external and internal grinding, centerless grinding, boned finishing, etc. in spite of different kinematics, unified approach will reduce the time to develop models and to ensure the required simulation fidelity.

Results and Discussion.

As an examplethere are results of systematic and random components of the machined surface roughness when turning. Cutting speed V = 1.32 m/s (n = 630 rev/min in workpiece diameter d = 40 mm). Depth of cut t = 0.5 mm. The feed was changed from S = 0.05 mm/rev up to S = 1 mm/rev. Processed material – steel 40X as received. Tool – straight-turning tool. Tool material – carbide material T15K6. The geometry of the tool with its spread is shown in Table 1.

Table 1 - The geometry of the tool when stochastic modeling

Option	**Range**	**Averagevalue**	**MSE**
The main angle in plan φ	60 ... 65 ^{0}	62.5 ^{0}	0,833
Wedgeangle ε	90 ± 1 ^{0}	90 ^{0}	0,167
Noseradius r	0.2...0.25 mm	0.225 mm	0.00833

Fig. 2 shows a graph of systematic component of the roughness parameter Ra depending on the feed, and Fig. 3 - the calculated distribution density of the roughness parameter when S = 0,1 mm/rev.

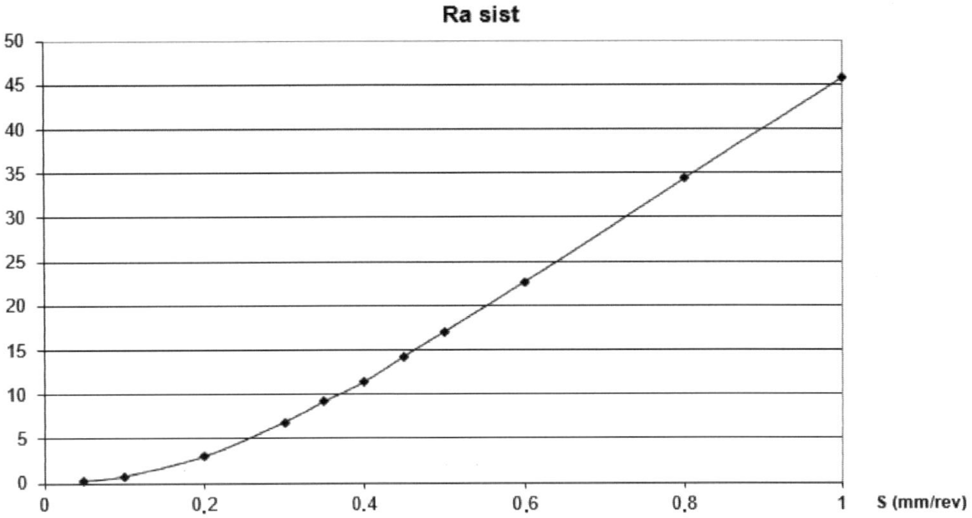

Figure 2 - The systematic component of Ra parameter (microns)

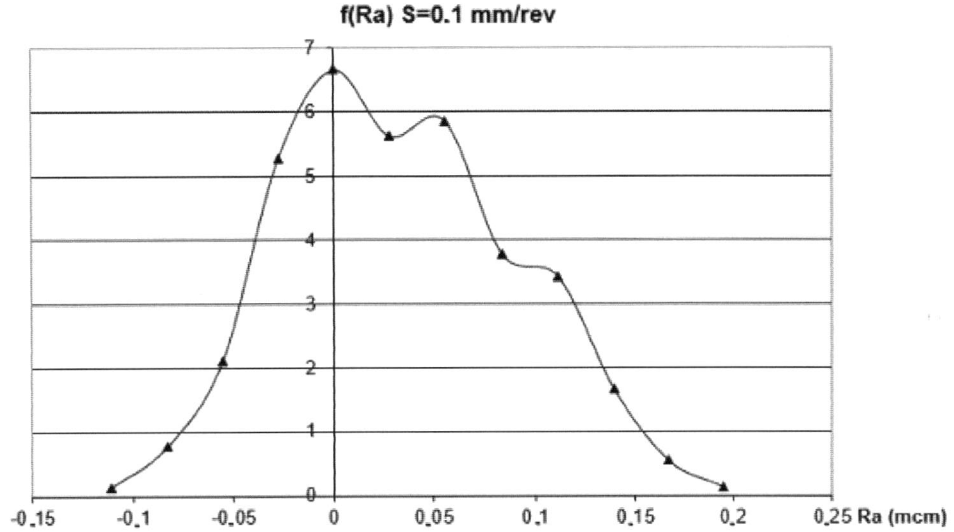

Figure 3 - The density distribution of the random component of Ra in S = 0.1 mm/rev

For analytical description of the density distribution a combined law was used by the Gaussian distribution:

$$f(Ra) = \frac{1}{\sqrt{2\pi}} \sum_{i=1}^{3} \frac{\alpha_i}{\sigma_i} \exp\left(-\frac{(Ra - Ra_{cpi})^2}{2\sigma_i^2} \right) \qquad (2)$$

where $\tilde{\alpha}_i$ weights: $\sum_{i=1}^{3} \alpha_i = 1$.

Conclusions

These models will allow solving a wide range of problems associated with the design of mechanical processing. To create the specific models of mechanical processing we need to mathematically describe and define the connection between the model and, primarily, the topography formation based on geometrical copy, oscillation deformation, plastic-elastic deformation and a random component.

References

[1] Suslov A.G. Scientific bases of mechanical engineering technology / A.G.Suslov, A.M.Dalsky - M.: Engineering, 2002. - 684 p.
[2] BaratzY.I. Optimizing the combination of regular microreliefs mating friction surfaces / Y.I.Baratz, R.K.Shaposhnik, V.M.Varchev // Engineering reporter. - 1992. - №5. - C. 18-20.
[3] ValetovV.A. Microgeometry surface and its performance characteristics / V.A.Valetov // Engineering reporter. - 1986. - №4. - C. 39-41.
[4] Karpunin M.G.Life cycle and efficiency of machinery / M.G.Karpunin, Y.G.Lyubinetsky, B.I.Maidanchik. - M .: Engineering, 1989. - 312 p.
[5] Leonov S.L. Basics of precision forming simulated technologies / S.L.Leonov, A.T. Zinoviev - Barnaul: Altai State Technical University Publishing House, 2006. - 198 p.
[6] Leonov S.L. Machining operations: Textbook / S.L.Leonov, E.Y.Tatarkin, Y.V. Fedorov - Barnaul: Altai State Technical University Publishing House, 2003. - 104 p.
[7] Kozlov V.I. Analysis of the relative motionimpact in edge tool deterioration/ V.I.Kozlov // STIN. - 2008. - №1. - C. 6-14.
[8] Kudinov V.A.Machines dynamics / V.A.Kudinov - M .: Engineering, 1967. - 360 p.

Computations of Accumulated Deformations and Depletion of Plasticity Reserve Held by the Metal in the Surface Layer during Orthogonal Cutting

V Yu Blumenstein [1,a], **A Ferranti** [2,b]

[1] 28 Vesennyaya str., Kemerovo, 650000, Russia, T.F. Gorbachev Kuzbass State Technical University
[2] Unit B, 6th Floor, Success Commercial Building, 251 Hennessy Road, Wanchai, Hong Kong, Big Wave (Hong Kong) Limited

[a] e-mail: blumenstein.vu@gmail.ru, [b] e-mail: alinaferranti@gmail.com

Abstract. The problem relating to the mechanics of orthogonal cutting with the deformation site, which is located within the contact area ahead of, below and behind the cutting tool was provided a solution. The above problem is crucial as it involves developing the mechanics of technological inheritance, which emphasizes the need for the operation-by-operation computations of the metal deformed state. The plane deformation (strain) model was adopted for the treatment with a cutter having a rounded cutting edge. Stresses, rates and deformations along the lines of current, which form the strengthened surface layer of the treated part, were computed. The patterns determining deformation accumulation and metal plasticity reserve depletion during orthogonal cutting were exposed.

Introduction

One of the crucial challenges of contemporary machine engineering is to make sure using processing methods that machine parts have a longer service life. The service life of machine parts is, in many ways, determined by the behavior of the surface layer. Its parameters are formed throughout the entire design process. When designing a route for strengthening machining and when assessing the service life of a machine part, allowance for technological inheritance (TI) should be provided. This means exposing and applying the functional dependencies between the parameters of the surface behavior and performance parameters. That, in its turn, requires an analysis of that behavior initiation at all stages of the life cycle of such a part.

The above problem is crucial as it involves developing the mechanics of technological inheritance, which emphasizes the need for the operation-by-operation computations of the metal deformed state [1-2]. What is important is to study the stages of the life cycle "on-line", i.e. to study the process of technological inheritance but not to assess the "frozen" result of stressing.

In order to achieve that, we need to know how the surface layer is formed and how to calculate the accumulated properties. The conventional parameters of strengthening (hardening) for the solution of such a problem are not acceptable. To do that, one needs to know the physical patterns of surface layer formation, which are assessed by means of deformation parameters.

It is a well-established fact that during cutting plastic deformation goes below the cutting edge of the tool to a specific depth.

N.N. Zorev, in his mechanical model of the cutting process, focuses on the plastic flow area - a deformation site - not only ahead of the cutter but also below the cutting surface (fig. 1) [3]. Such a model confirms and describes the well-known facts relating to strengthening of the machined surface layer of the metal as deep as 0.1-4.5 mm [4]. C. Cassin and G. Boothroyd emphasize the presence of five active areas during metal cutting, including the machined surface layer area (fig. 2) [5]. The authors pinpoint that secondary deformation is the distortion, which arose from the tool affecting the workpiece and the friction between them.

R.N. Roth and P.L.B. Oxley, based on the observations, suggested a kinematic mesh of slip lines, which confirmed that a deformation site has final dimensions (fig. 3) [6]. Despite an obvious lack of some boundary conditions on the free surface and their satisfaction, the above model gives a quality description of the real time process.

M.M. Barash and W.J. Schoech, by assuming that plane strain is oriented along the cutting velocity and by formulating a problem from the standpoint of elastic plastic deformation, come to statically possible stress fields (fig. 4) [7]. Such a model describes a zone of plastic deformation in the layer beneath metal surface. The authors believe that "if the velocity is checked up to the neighborhood of the lowest point of the fan, conditions obtain which are similar to those found in the Usui and Hoshi field". Such field satisfies velocity conditions.

The depth of the plastically deformed layer (strengthening depth) was analytically defined and also tested experimentally by measuring the hardness, by determining the layer with residual stresses and the depth of side spread, occurring underneath the surface treated by cutting. The work shows the consistency of the plane strain model, which further received widespread acceptance. We need to highlight that in this model the cutter is acute-angled, though, in actuality, we deal with the rounded radius of any cutting edge. More recent works regarding plastic deformation established that the depth of the plastically deformed layer is different from the depth of residual stress distribution.

Ocuschimo K, Kakino Y., by using finite element method (FEM), resolved the problem relating to the determination of the residual stress state affected by heat and force in the conditions of orthogonal cutting [8].

Iwata K., Osakada K., Terasaka Y. applied the FEM to a rigid plastic body during plane and stationary orthogonal cutting [9]. The authors believe that the stress and strain history affects plastic failure, therefore the calculations to determine such a failure were made along the lines of current. The failure criterion (the authors notation) was seen as follows:

$$\int_0^{\overline{\varepsilon_f}} \frac{\sigma^*}{\overline{\sigma}} d\overline{\varepsilon} = c, \tag{1}$$

where $\overline{\varepsilon_f}$ – equivalent deformation by the moment of failure; σ^* – maximum normal stress; $\overline{\sigma}$ – yielding stress. The specimen made from medium carbon steel (0.45% C) was treated with a cutter from high-speed steel with a front angle of $\gamma = (10-30)°$ as deep as t=(0.02–0.1) mm . The study showed that deformation rate has its largest value of $\overline{\varepsilon}$=30·10^4 c^{-1} around where the cutting edge is; the largest value of equivalent deformation on the front contact and machined surface is $\overline{\varepsilon}$=1.6. It was established that the nature of metal flow is significantly affected by prior metal strengthening, rounded radius of the cutting edge and other process aspects.

Several works give findings relating to the study of metal failure during cutting. P.D. Bespakhotniy and Yu.V. Fedorov, by conducting analytical and experimental studies, showed that in some cases chips form together with the occurrence of a visible crack ahead of the wedge [10] (fig. 5).

V.A. Timotshenko and E.V. Goldysh studied the patterns of material plastic flow and its failure in the chip formation site during orthogonal cutting [11-12]. By using the method of slip lines, the authors calculated the incremental deformation when a material particle travels across shear line ACF as well as a change in the indicator of plasticity reserve depletion on this line (Fig. 6). A linear model to calculate plasticity reserve depletion was used and the non-monotony of the cutting process wasn't

IOP Conf. Series: Materials Science and Engineering **126** (2016) 012010 doi:10.1088/1757-899X/126/1/012010

analyzed, which, in its turn, did not allow for the calculation of the strengthening of the surface layer being formed.

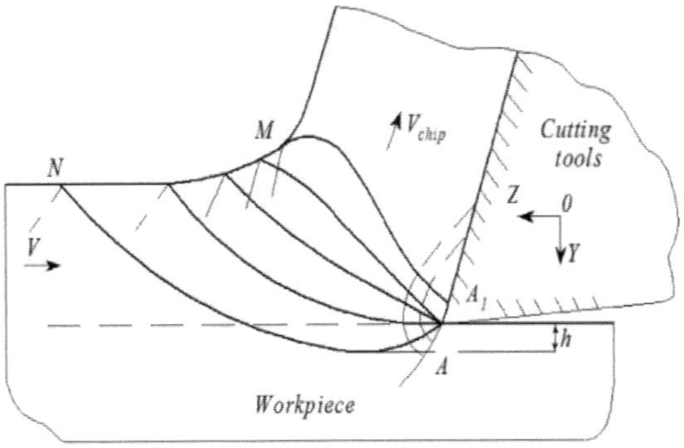

Fig. 1. Mechanical model of cutting: h – strengthening depth

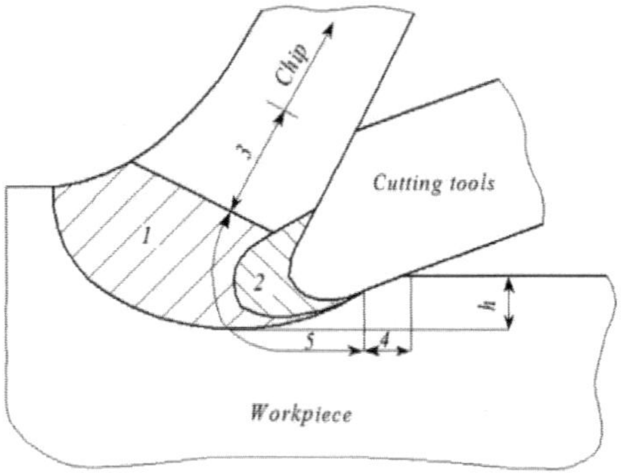

Fig. 2. Cutting schematic: 1 – primary deformation zone; 2 – secondary deformation zone; 3 – surface of the chip contact; 4 – surface of the contact between a workpiece and a cutter; 5 – cutter's edge

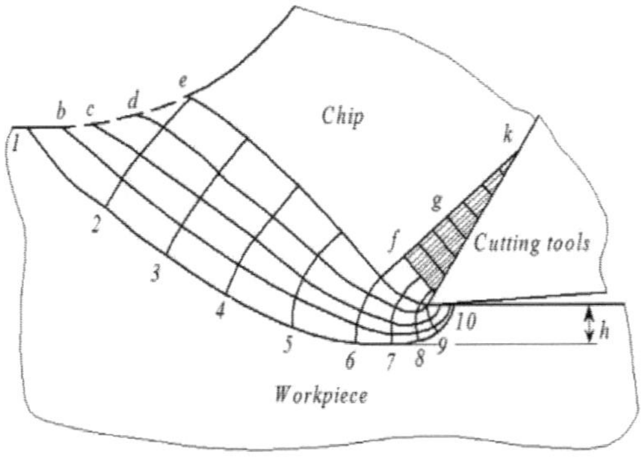

Fig. 3. Mesh of slip lines according to the observations

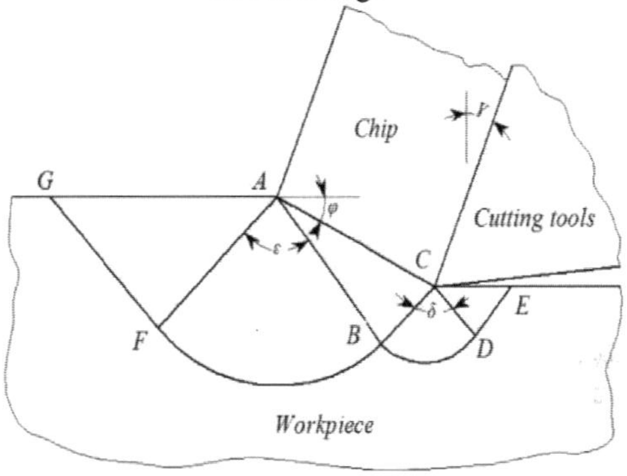

Fig. 4. Idealized cutting model

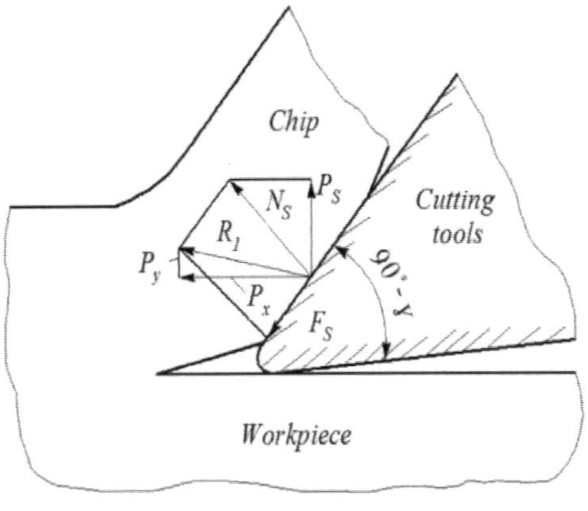

Fig. 5. Crack growth ahead of the wedge

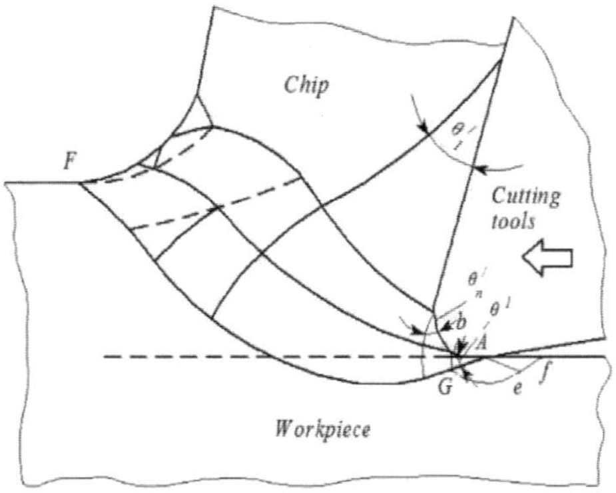

Fig. 6. Field of slip lines at the stage of the settled cutting

S.G. Melikhov applied the plastic flow theory of radiation-hardened material [13]. The study was conducted in the conditions of orthogonal cutting, the work material was plastic heat-resistant alloy XH75MBU with the following mechanical properties: point of yielding R_B=550-650 MPa; ultimate resistance R_m=950-1180 MPa; Brinell hardness HB 255-321. Cutting was performed with a cutter having a cutting edge with a rounded radius ρ=0.05 mm, front angle γ=35°, back angle α=8°, cutting edge inclination λ=0. The cutting depth was t=0.35 mm, feed was S=0.032 m/min. The selected mode and specimen dimensions 25x25x6 mm provided adherence to the conditions of plane deformation, stationary and quite slow metal flow, insignificant impact of temperatures, flow chips and an absence of buildup on the edge.

By analyzing the kinematics of metal flow and by using the slip line method the author received the stress field in the combination with the velocity field. The solution of this problem used the slip line method on the basis of the theory of rigid plastic and radiation-hardened body (fig. 7). What is important is that in the above picture of flow the current lines, beginning with some final one, which travels through the cutter top and lower, look wavy. As a result, the velocity of metal particles along the current lines was analyzed and the presence of some dividing point on the cutter top, where the velocity is 0, was proved. In this dividing point, the metal flows are divided into those turning into chips, those going down the cutter and those going into the work surface. The value of accumulated plastic deformation for the state of plane deformation was seen through the components of deformation rates.

O.I. Mizhiritskiy, A.A. Bogatov and V.F. Igoshin used the method of characteristics to solve the boundary problem for the incompressible and ideally-plastic medium, with the rounded cutting edge being approximated with three straight-line portions for the sake of simplicity [14] (fig. 7). The paths of metal particles within the deformation site were determined, stress and strain indicators, including the accumulated degree of shearing deformation Λ, were determined along such paths. The non-monotony of deformation development was assessed as to how angle ϕ was altered, which characterizes the rotation of primary stress axes against the material particle. The metal damage (degree of plasticity reserve depletion) was computed as follows [15]:

$$\omega = \sum_{i=1}^{n} \frac{a\Lambda^{a-1}}{\Lambda_p^a} d\Lambda , \qquad (2)$$

where Λ_i is a degree of the shearing deformation on the -m segment of quasimonotonous deformation; Λ_p and α are plastic property indicators of the metal under treatment, which depend on the relation of average normal stress to the intensity of shear stresses σ/T, Lode parameter μ_σ, the intensity of shearing deformation rate H, the temperature Θ, the structure parameter $\overline{x_i}$ and the contents of chemical elements in the alloy c_j. The indicator α factors the non-linear damage accumulation in the model (2). As a result, the authors received the distribution of the shearing deformation degree Λ and the damage ω as to the depth of the strengthened layer and the developed computations of damage were used to define rational modes for the turning of hot-rolled pipes prior to their coming plastic deformation.

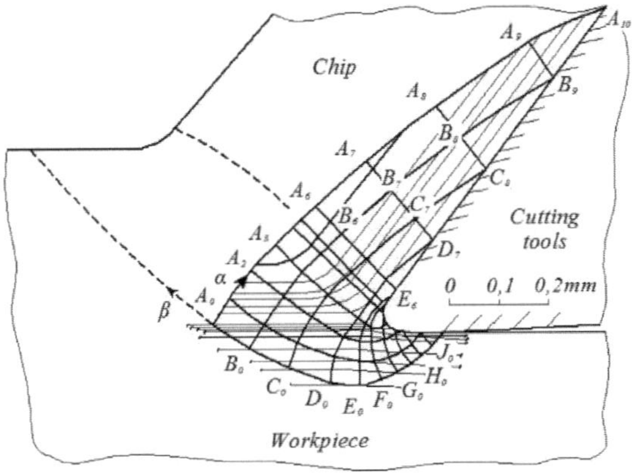

Fig. 7. Field of slip lines and current lines within the deformation site

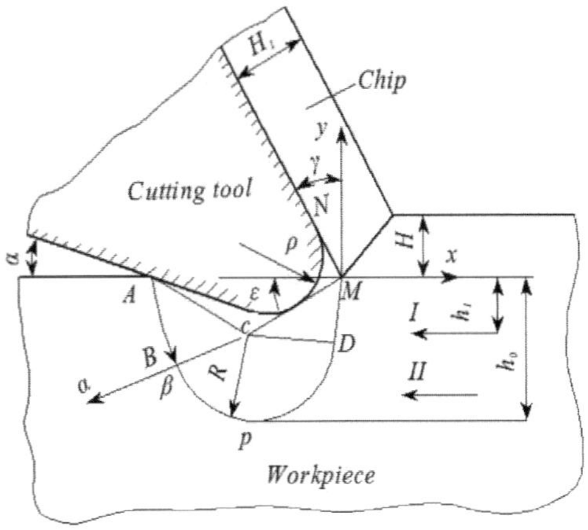

Fig. 8. Schematic of the deformation site during the turning operation

Results and Discussion

The analysis of the findings, including by means of grade grids, laid the basis for the analysis of metal flow patterns within the plastic deformation site (DS). Also, the schematic of the settled (stationary) orthogonal cutting for plastic steel was proposed [16] (fig. 9).

When the cutter with a cutting edge having a rounded radius ρ moves by the value of feed within the contact area, the deformation site occurs, where metal flows, deformation accumulates and metal plasticity reserve depletes along current lines (for instance, 1, 2, 3).

According to the flow pattern, deformation site *KLMCDEFGAK* is seen to comprise two areas: higher and lower than some current line 1, which overlaps with line *ABC*.

The DS boundaries are described with a set of lines:

Curved line KL describes non-contact chip edge;

- Curved line LM, which is the end line of metal plastic flow; point M stands for the end of plastic contact of the chip with the cutting tool and point N stands for the end of elastic contact of the chip with the cutter;
- Curved line KAG is the starting line of metal plastic flow (front boundary of deformation site). To the left of this line the metal is in its elastic state, to the right – in its plastic flow state. Point G stands for the inmost depth of the plastic deformation growth;
- Curved line GF is the end line of plastic metal flow (back boundary of deformation site);
- Curved wavelike line ABC – critical current line, which separates metal flows into those turning into the chip and those going down the tool. Point A stands for the start of metal plastic rise, the curvature of this line alters in point B, point C stands for the highest rise of the metal in the form of a plastic wave preceding the cutter;
- Curved line MCDE describes the contour line of the surface of the cutter contact with the part and the chip. Point D stands for the largest depth of the cutter penetration;
- Curved line EF describes the back non-contact boundary; point E is the point of separation of the cutter from the treated surface.

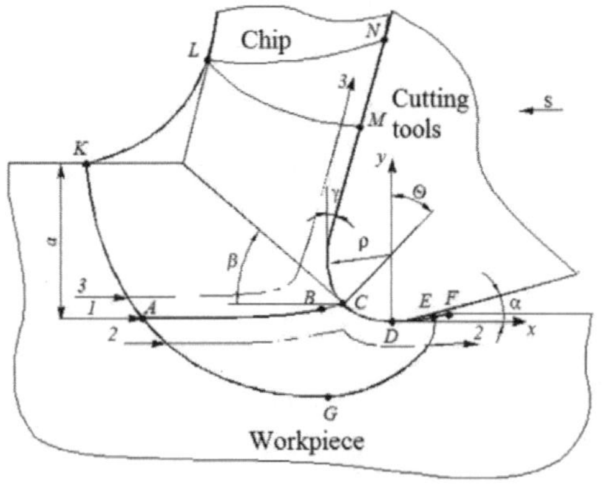

Fig. 9. Schematic of the deformation site in the conditions of orthogonal cutting [15]: α – rear angle; γ – front angle; ρ – tool edge radius; Θ – indentation angle; a – cutting depth; S – feed

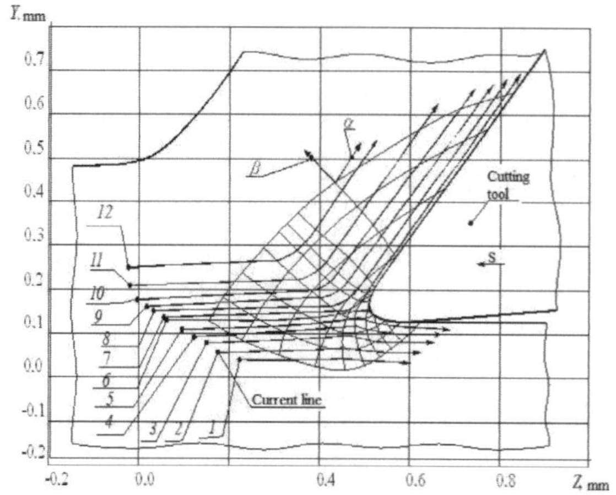

Fig. 10. Computational diagram of the deformation site: critical point z=0.49 mm

In the authors' opinion, depending on the stressing diagram, the accumulated deformation and the degree of plasticity reserve depletion metal flows may split in point *A*, along current line *ABC* or in point *C*, which will generate miscellaneous kinds of chips. Complete plasticity reserve depletion around point A, which is characteristic of low-ductility metals, causes the formation of free front and non-contact surface ABC and segmental chips. The accumulation of maximum permissible deformations in critical point C causes discontinuity to occur and discontinuous chips to be formed.

Such an approach and the assessment of DS formation allowed for the formulation of initial and boundary conditions, for the resolution of the problem relating to the mechanics of orthogonal cutting and for the computation along the current lines of the accumulated deformations and metal plasticity reserve in the strengthened surface layer.

This study is based on the model suggested by S.G. Melikhov, which reflects the real cutting process to the fullest extent [13]. The authors constructed the current lines and, according to the initial data, made the computations of the components and invariants of stress tensors, deformations, deformation rates and other deformation parameters in some points on the current lines, including shear stress intensity T; normal stress intensity σ_i; average normal stress σ; indicator of stress state $\Pi = T/\sigma$; deformation intensity ε_i; intensity of deformation rates ξ_i; intensity of shearing deformation rates $H = \sqrt{3}\xi_i$ (fig. 10).

The degree of shearing deformation was determined as follows:

$$\Lambda = \int_0^t H d\tau , \quad (3)$$

where the time interval for the passing by the reviewed particle from the occurrence of the plastic deformation $\tau = 0$ till the completion of the plastic flow $\tau = t$ is under discussion

For this problem the flow curve looked as follows:

$$T = 34.6 + 65.8\Lambda^{0.25} \quad (4)$$

The degree of shearing deformation Λ_p (see also formula (3)), accumulated by the metal particle by the moment of macrofailure in the permanent conditions of deformation temperature and rate and unchangeable parameters of the stress state, is regarded as a measure of plasticity as follows:

$$\Lambda_p = 10.1\exp(-1.2\Pi) \quad (5)$$

The degree of plasticity reserve depletion was assessed as follows [17]:

$$\Psi = \Psi_1 + \Psi_2 = \Psi_1 + (\Psi_{21} + \Psi_{22}) = n\varphi_0 \int_0^{\Lambda_k} \Lambda_i^{n-1} d\Lambda \left(\int_0^{\Lambda_k} \frac{d\Lambda}{\Lambda_p} - \varphi_0 \int_0^{\Lambda_k} \Lambda_p^{n-1} d\Lambda \right), \tag{6}$$

Ψ_1 – component dependent on flow stress or on accumulated deformation; Ψ_2 – component dependent on metal plasticity with $\Pi = const$; Λ and Λ_p – accumulated and maximum permissible degree of shearing deformation with a specific stress state indicator Π; n – strain-hardening coefficient; φ_0 – coefficient determined by plasticity tests. In the unstrengthened metal $\Psi = 0$, when plasticity reserve is completely depleted, $\Psi = 1$.

Computations of the stress and strain state (SSS) are made for current lines 1-6, which are forming the surface layer of the machined part, and current lines 7-12, along which the metal plastic flow in the chip occurs. The SSS for the critical 6th line of current is of utmost interest as it divides the metal flows into those turning into the chip and those going down the tool. It is crucial to know what the conditions are that determine the formation of the surface layer and how the detachable chip affects such formation.

It is established that the distribution of stresses and deformations is complex by nature as within the deformation site, along the 6th and the neighboring 5th and 7th current lines, we observed some periodic behavior in the distribution of the SSS components.

The stress component σ_y along the 6th line of current, at the beginning, has a negative value $\sigma_y = -252$ MPa, then it changes sign by reaching the maximum positive value $\sigma_y = 1317$ MPa in the point with the abscissa z=0.45 mm, then it decreases by moving into the negative zone, and where critical point with the abscissa z=0.49 mm is and along the abscissa axe has a tendency to increase (fig. 11). The distribution of the component σ_z is even more complex by nature. At the beginning $\sigma_z = -1167$ MPa, then its absolute value goes down and around point C it dramatically changes. The component value grows till it reaches a maximum positive value $\sigma_z = -1640$ MPa, then again there's a dramatic change, its value goes down and goes back up when it leaves the deformation site. The shearing component σ_{zy} remains negative all the time, however also goes through three changes of its polarity in the distribution along the discussed 6th line of current.

As a result, the distribution of average normal stress is quite complex along the 6th line of current, especially around the critical point. It is established that the intensity of shearing stresses reaches its maximum value T=1262 MPa also around that critical point. The indicator of the stress state along the 6th line of current alters twice. It is clearly seen that the stress state around point C is close to the state of the pure shear (fig. 12).

The deformation rates alter dramatically around the critical point (fig. 13-14). Specifically, deformation rate component ξ_z, by remaining negative, goes down till it reaches the value of $\xi_z = -12.6$ c^{-1}, then there is a radical change and it increases with z=0.49 mm till $\xi_z = 22.0$ c^{-1}, then it changes again and it goes down till $\xi_z = -1.3$ c^{-1}. The normal component of the deformation rate alters in a similar way. Around the critical point the distribution of the shear velocity η_{xy} alters twice (fig. 15).

The analysis of the type of the stress state in the distribution of stress components, deformation rates and intensity along current lines 1 through 6 showed that there are three areas of quasimonotonous deformation (fig. 9): the first stretches from the DS entry point A to the point of the metal highest rise B; the second – from point B to critical point C; the third – the segment under the tool CDE. Thus, along the 6th line, first of all, around the critical point, there are favorable conditions for the division of the metal flows: a dramatic change in the distribution of normal and shearing stress components, shift and deformation rates – all that leads to the dramatic increase in the intensity values of shearing deformation rates. It is shown that the intensity of shearing deformation rates along the 6th current line reaches its maximum value H=0.75 c^{-1} also around the critical point (fig. 16).

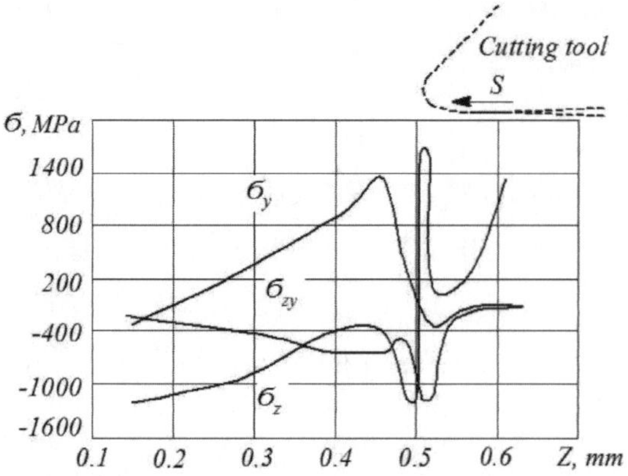

Fig. 11. Distribution of stress components along the 6th current line

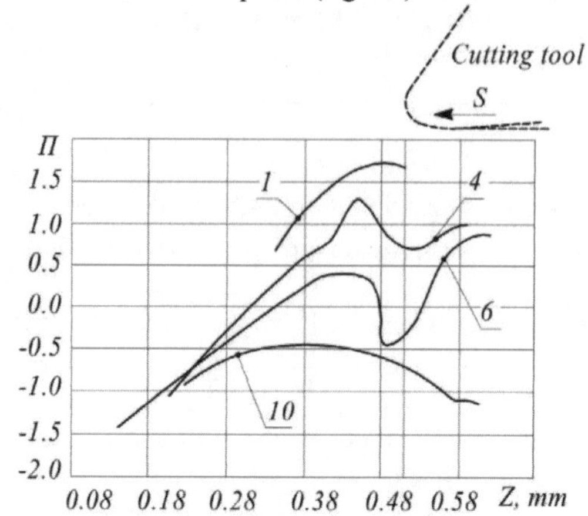

Fig. 12. Distribution of stress state indicator along some current lines

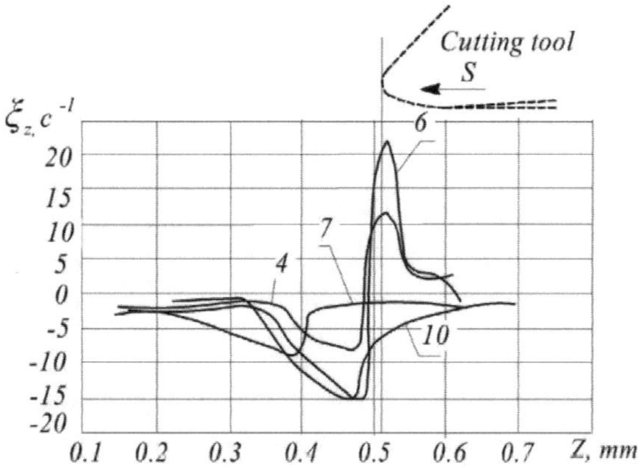

Fig. 13. Distribution of deformation rate component ξ_z along some current lines

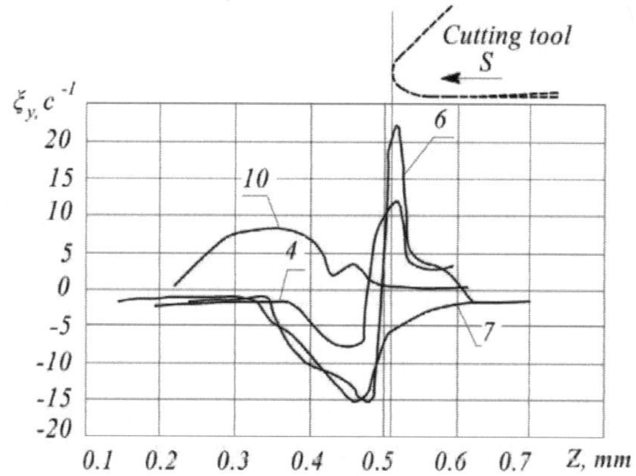

Fig. 14. Distribution of deformation rate component ξ_y along some current lines

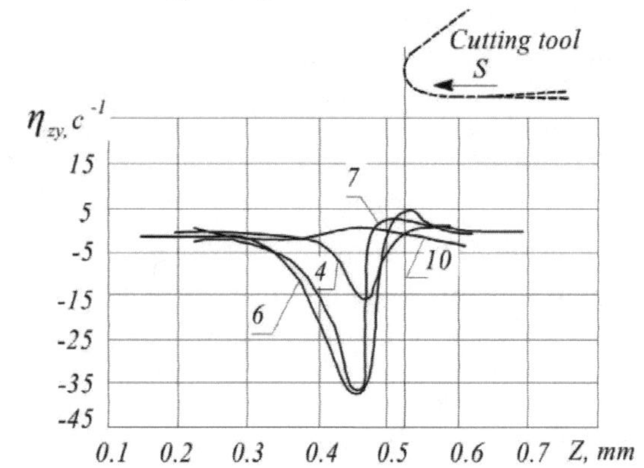

Fig. 15. Distribution of shear rates η_{xy} along some current lines

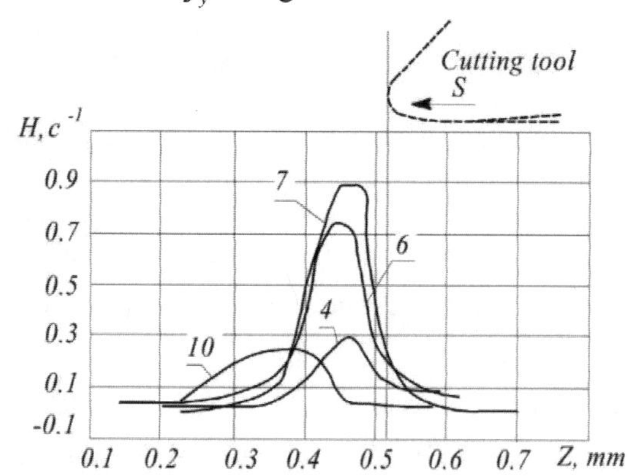

Fig. 16. Distribution of the intensity of shear deformation rates H along some current lines

The computations that we made helped receive the distribution of the degree of shear deformation and the degree of plasticity reserve depletion along the depth of the strengthened surface layer on the back boundary of the deformation site (fig. 17-18).

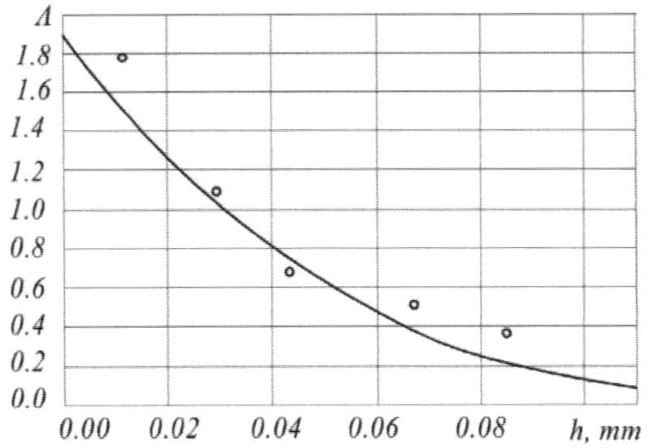

Fig. 17. Distribution of the degree of shear deformation along the depth of the surface layer on the back boundary of the deformation site

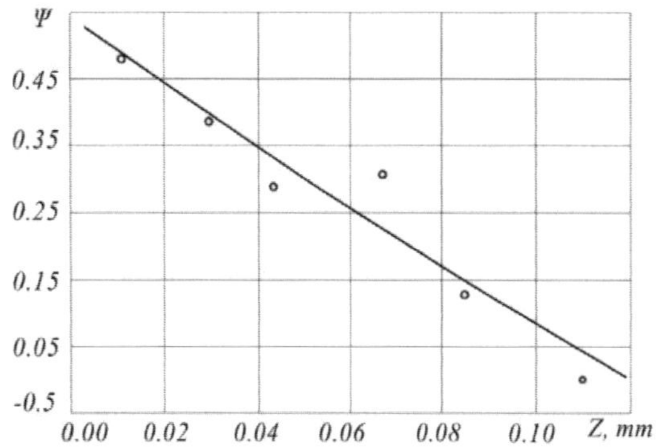

Fig. 18. Distribution of the degree of plasticity reserve depletion along the depth of the surface layer on the back boundary of the deformation site

What is very characteristic of this distribution is that 63% of all the deformation is accumulated in the thin strengthened surface layer as deep as 0 through 0.04 mm, the remaining deformation, which is 37% is in the layer as deep as 0.04 through 0.11mm.

The degree of metal plasticity reserve depletion in the strengthened surface layer was around $\Psi=0.5$. Therefore, the residue reserve of plasticity $\Psi=1\text{-}0.5\text{-}0.5$ can be distributed during the subsequent treatment and operational stressing (loading).

Thus, the computations did not record any complete depletion of metal plasticity reserve, and along the 6th line of current, including around critical point C, the plastic division of metal flows into the chip and down the tool takes place without any failure. That matches the real picture of the metal flow [13]. One can suggest that this kind of plasticity reserve depletion corresponds to the formation of flow chip.

In the authors' opinion, formation of this or that type of chip is governed by the patterns of plasticity reserve depletion. The plasticity reduction and the "tougher" stressing diagram lead to an earlier and complete plasticity reserve depletion on the critical line of current and cause the shearing chip to form. In this case, the deformation site below the critical line of current is formed, to a greater extent, when free front and non-contact surface becomes available (fig. 5) [10].

Conclusion

1. The model for the surface layer formation during orthogonal cutting with the deformation site, which is located where the cutter contacts the part, was proposed. It is shown that the metal flow occurs along the lines of current, some of which form the strengthened surface layer of the treated part and the others form the chip.
2. The critical line of current, which divides the metal flows into the chip and those going down the tool was detected.
3. The computations for the metal stress and strain state were made with the exposure of the patterns of deformation accumulation and the degree of plasticity reserve depletion of the metal belonging to the strengthened surface layer of the part.
4. The reasons why the metal flows divide along the critical line of current and different types of chip form were given an explanation from the standpoint of the mechanics of deformation.
5. The solution that we obtained allowed for the formulation of initial and boundary conditions for the resolution of the problems relating to subsequent treatment and performance operations that such a part needs to go through, as well as for the study of the patterns and the production of concepts regarding technological inheritance by means of the unified terms and interpretations of deformable solid mechanics.

IOP Conf. Series: Materials Science and Engineering **126** (2016) 012010 doi:10.1088/1757-899X/126/1/012010

References

[1] Blumenstein V.Yu., Mechanics of Technological Inheritance during Treatment and Operation of Machine Parts / V.Yu. Blumenstein, V.M. Smelyanskiy, Mashinostroyeniye-1, Moscow, 2007, 400 pp.

[2] Blumenstein V.Yu., Mechanics of Technological Inheritance. International Journal for Science and Innovations for the Industry. Innovations in Discrete Productions Issue. Year III, Issue 2/2015, Sofia. ISSN 1314-8907. 2015, №2, pp 18-21.

[3] Zorev N.N., Problems of Metal Cutting Mechanics, Moscow: Mashgiz, 1956, 338 pp.

[4] Isaev A.I., Surface Layer Formation during the Treatment of Metals with Cutting //Collection of TZNIITMASH, Moscow: Mashgiz, 1950, Issue 33, 358 pp.

[5] Cassin C., Boothroyd G. Lubricating Action of Cutting Fluids//J. Mech. Eng. Sci. -1965. - 7. - 67.

[6] Roth R.N., Oxley P.L.B. Slip Line Field Analysis for Orthogonal Machining Based upon Experimental Flow Fields//J. Mech. Eng. Sci. -1972. -14. - 85.

[7] Barash M.M., Schoech W.J. A Semi-Analytical Model of the Residual Stress Zone in Orthogonal Machining // Advances Mach. Tool Design and Res. - 1970. -p. 603-613.

[8] Ocuschimo K, Kakino Y. A Study on the Residual Stress Produced by Metal Cutting. - Mem. For Eng. – Kyoto Univ., 1972. – V.34. – P. 234 - 248.

[9] Iwata K., Osakada K., Terasaka Y. Process Modelling of Orthogonal Cutting by the Rigid-Plastic Finite Element Method /Collection of the American Society of Mechanical Engineers. – 1984. – Volume 106. - Issue 2. – P. 24 - 31.

[10] Bespakhotniy P.D., Fedorov Yu.V., Study of Cutting from the Standpoint of the Failure Theory//Applied Problems of Physics and Material Failure. Materials of the meeting (Research Institute of Technology and Production Organization), Moscow: NTO Mashprom, 1974, P. 54-62.

[11] Timotshenko V.A., Goldysh E.V., Formation of the Part's Surface during Orthogonal Cutting //STIN. – 1995. - № . 7 – P. 16 - 18.

[12] Timotshenko V.A., Goldysh E.V., Connection between Cutting Conditions and Maximum Permissible Plastic Deformation of the Metal within the Chip Formation Site//STIN. – 1996. - № 11. – P. 14-16.

[13] Melikhov S.G., Method for the Calculation of the Metal Stress and Strain State during Cutting Based on the Theory of the Plastic Flow of Nonhomogeneous Body: Thesis for a Candidate of Sciences degree - M., 1971. - 172 p.: illustr. - References: P. 16-171.

[14] Mizhiritskiy O.I., Bogatov A.A., Igoshin V.F., Determining the Damage of the Near-surface Layer of the Metal during the Turning Operation// Treatment of Metals with Pressure. - Sverdlovsk, 1987, № 14. - P. 102-113.

[15] Bogatov A.A., Mizhiritskiy O.I., Smirnov S.V., Life Time of Metal Ductility during Treatment with Pressure. - Moscow: Mettalurgy, 1984. - 144 p.

[16] V.Yu. Blumenstein, Technological Inheritance Mechanics. The Depletion of Plasticity Reserve during Cutting. International Conference "Fundamental and Applied Technological Problems of Machine Building" "Technology - 2000". 28-30 September. Oryol, Russia. – P. 111-114.

[17] Yu.K. Filippov, Evaluation Criterion for the Quality of Parts Produced by Cold Forging//Forging and Stamping Production, 1999, № 2, P. 3-9.

IOP Conf. Series: Materials Science and Engineering **126** (2016) 012011 doi:10.1088/1757-899X/126/1/012011

Pallet Optimization of the Heavy Rotary Table Load Carrying System

V G Atapin [*], **A A Bataev**

Faculty of Mechanical Engineering and Technologies, Novosibirsk State Technical University, 20 Prospekt K. Marksa, Novosibirsk, 630073, Russia

*E-mail: teormech@ngs.ru

Abstract. The pallet optimization of the heavy rotary table load-carrying system, which is a part of the multi-purpose machine, is considered in terms of the deterministic and probabilistic models. As a result of optimum design in case of the deterministic model the mass of the pallet is reduced by 35.5 % in comparison with a serial model. The evaluation of the influence of optimization problem limitations on design variables confirms the importance of rigidity criterion in relation to other criteria. Calculation for probabilistic model allows reducing the mass of the construction by 27 % in comparison with the deterministic model. Considering a work piece rigidity on the basis of a conventional work piece of the minimum rigidity (without stiffening ribs etc.) leads to reducing of the pallet mass by 22.3 % in comparison with the deterministic model.

1. Introduction

Over the past decades, column-and-knee-type machines were widely used due to their ability to process the work piece from all sides. On the first modifications of the machines, the work piece rotation was carried out with the help of crane, the procedure taking from 0.5 to 3 hours. In most cases, the processing of the work piece surface on one side takes less than a working shift duration, which leads to intensive crane operation and working efficiency reduction. The application of rotary tables eliminates the mentioned problems. The work piece is placed onto the rotating part of the table (pallet) and the desired position precision in relation to the machine is achieved by the movement of the rotating part.

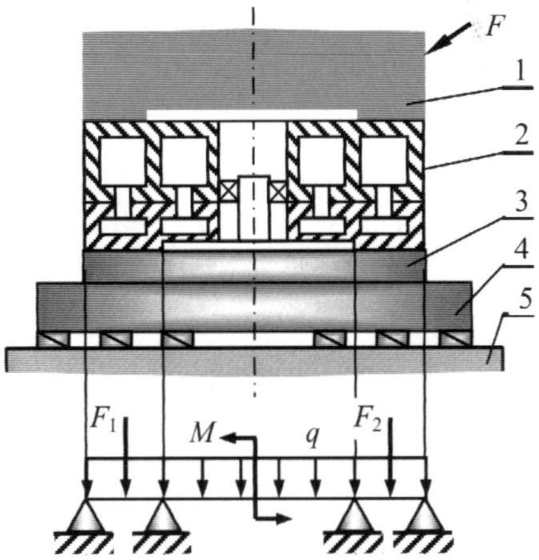

Figure 1. Pallet calculation scheme: 1 – work piece, 2 – pallet, 3 – rotating table, 4 – bed, 5 – foundation

Heavy column-and-knee-type machines [1] are often used for the processing of the mechanisms basic parts. The outer dimensions of these parts have greatly changed with the advances in technology. Originally, the dimensions of the work pieces were no bigger than a flatcar and the geometrical precision of the surfaces and their relationship was achieved at the final assembly stage by the manual fitting. This imposed certain limits to the acceptable table body deformation; e.g. multisupport beam (Figure 1) was used as an analytical model for the body of the rotating part (pallet) with strength of materials methods for the rigidity calculation.

With the growth of the mechanisms operating parameters, such as capacity, pressure, temperature and load, there was an increase in outer dimensions and basic parts production precision, which in turn leads to the enlargement of the rotating tables clamping surface and work piece carrying capacity reducing at the same time the acceptable deformation of the table body caused by the weight of the processed work pieces. Under these conditions the necessity and importance of rotating tables strength, rigidity and other performance criteria validity is increasing. The updated requirements to the rotating table construction can be partially satisfied by the designer decisions aimed at the symmetrical load distribution provision, supporting plane dimensional growth, unsupported length reduction, but mostly at the expense of manifold increase in the cross-section moment of inertia, i.e. outer dimensions and mass.

2. The statement of the problem

Spatial repositioning of a spot on the processed work piece, which is placed on the rotating table, in the process of its reorientation within the operating zone depends on the table rigidity, work piece rigidity, the mass of the processed work piece and its center of gravity position in relation to the table pivot axis as well as on the magnitude and direction of the cutting force. The mentioned parameters with the exception of the table rigidity are the initial data to be considered while designing. They are in part defined by the work pieces nomenclature, in part by the specifications of the machine and cutting tools. The table rigidity conditioned by the pallet rigidity, bed and foundation bodies and joints contact rigidity is to be determined during machine designing.

The implemented designing decisions in the table load carrying arrangement allow the bed and the foundation to experience mainly the compressive deformation under the external load. Therefore, the height of the bed and foundation is set to minimal allowed value for the design and technology reasons. The pallet mostly undergoes the bending deformation. Since the bending displacement is considerably greater than the compressive one and considering the fact that the load carrying system rigidity is primarily conditioned by the pallet rigidity, the prior attention will be given to the design of the pallet as the most deformable structural element.

The serial pallet (Figure 2) is a three-dimensional thin-walled rectangular formation of cellular structure with the dimensions $L = 5.6$ m, $B = 3.6$ m, $H = 0.8$ m [2]. There are lengthwise and crosswise stiffening ribs of rectangular cross section along the pallet lower edge. The pallet body rests on the bed guide track of ring cross section (outer diameter 3.6 m).

The pallet analytical model is based on the following propositions:

1. The pallets body is modeled with the help of plate-like rectangular finite element and rod finite element (ribs).
2. The pallet rests on rigid circular guide track of the table bed.
3. The calculated load is the pallet weight and the processed work piece weight (2 MN). The cutting forces due to their relative insignificance in comparison with the mentioned loads are not taken into account because during the finishing front milling work the maximum component of cutting force is 3.0 kN.
4. The processed work piece is supposed to be placed on the technological base, which coincides with the pallet corner zones. The external load F from the weights of the work piece and pallet is in extreme event characterized by forces F_i ($i = 1, \ldots, 4$), applied at the pallet corner points. The distribution of the load from the work piece weight in the pallet corner points is calculated by the strength of materials methods:

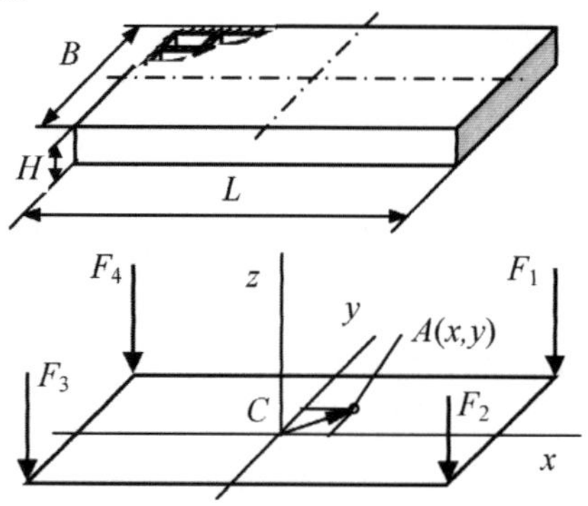

Figure 2. Pallet calculation scheme

$$F_i = (1/4)F\left[1 \pm x/(L/2) \pm y/(B/2)\right] \tag{1}$$

5. In general case the work piece center of gravity A (Figure 2) is moved in the plane xy relative to the table pivot axis by 1/20 of the length and 1/30 of the breadth of the pallet; this is the maximum eccentricity value found out by the analysis of the occurring large scale work pieces geometry.

3. Results and Discussion

a. Deterministic model

The coordinates of point A to apply the resultant load:

$$x = L/20 = 5.6/20 = 0.28\ \text{m},\ y = B/30 = 3.6/30 = 0.12\ \text{m}.$$

The forces $F_1, \ldots, F_4$ applied at the pallet corner points are defined with the help of the formula (1).

The problem of the pallet optimal design is stated as follows:

to minimize
$$\psi_0 = \rho\left(\sum_{i=1}^{k} V_i + \sum_{j=1}^{m} V_j\right) \tag{2}$$

limited by:

displacement	$\psi_1 = 1 - \delta / [\delta] \geq 0,$
stress	$\psi_2 = 1 - \sigma_{eqv} / \sigma_{allow} \geq 0,$
stability	$\psi_3 = 1 - n\sigma / \sigma_{cr} \geq 0,$
frequency	$\psi_4 = p_1 / [p_1] - 1 \geq 0$
design variables	$\psi_5 = V_i \geq 0, \quad i = 1, \ldots, k,$
	$\psi_6 = V_j \geq 0, \quad j = 1, \ldots, m,$

with k, m – being the number of plate-like rectangular and rod finite elements (FE); ρ – density of the material; V – the volume of the finite element; δ, $[\delta]$ – calculated and permissible relative deformation, rated in the direction perpendicular to the pallet plane; σ_{eqv}, σ_{allow} – equivalent stress and allowable stress; $n = 2$ – stability factor; σ, σ_{cr} – compressive stress and the critical compressive stress; p_1, $[p_1]$ = 12 Hz – calculated value and the lower limit (defined by the spindle rotating speed 500 min^{-1} with 30% resonance offset) of the first natural frequency.

The design variable are the thickness of the body wall – t_c and the thickness of the rib – t_p (rib breadth is constant). The pallet outer dimensions (length, breadth and height) are defined by the preliminary specifications and do not vary.

The main criterion to characterize the pallet rigidity is the pallet tilting angle because it directly influences the performance of hydrostatic guide track. The calculated pallet rigidity standard was introduced on the basis of this criterion, i.e. the relative vertical deformation $[\delta] = 2 \cdot 10^{-5}$ with guide track breadth being 1 m and oil lubricant layer thickness $4 \cdot 10^{-5}$ m.

The objective function of the problem (2) is the mass of the formation as firstly, the formation mass of several dozen tons is to be calculated, and secondly, certain criteria such as displacement rigidity and stress stability and some others can be assigned within permissible limits.

The problem (2) is solved by the penalty function method [3, 4] in the form

$$\varphi = \psi_0 / \psi_0^{\text{H}} + r\sum_{i=1}^{4}(1/\psi_i), \tag{3}$$

with ψ_0^{H} – the serial pallet original mass prior to optimization; r – small positive parameter. The problem is solved by the function (3) unconstrained minimization for the decreasing sequence of parameter r values with the help of Davidon-Fletcher-Powell method [4].

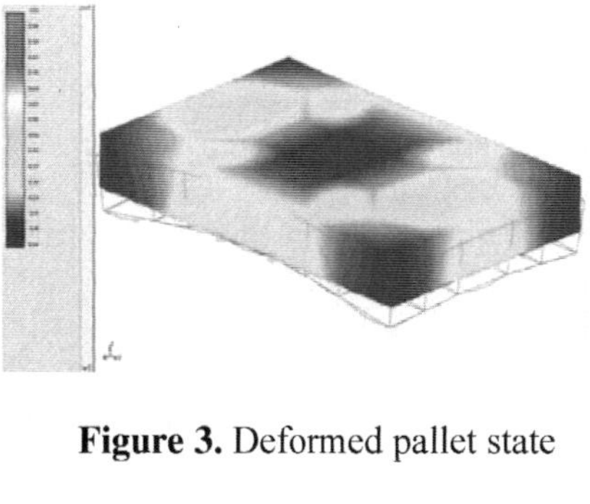

Figure 3. Deformed pallet state

Figure 3 demonstrates the deformation of the pallet, Table 1 presents the calculation results based only on Finite Elements Method (FEM) by the limited enumeration of possibilities as well as on FEM and optimization methods jointly. As a result of optimization designing the pallets mass was reduced by 35.5% in comparison with the serial version, which virtually coincides with the results achieved by FEM only. The difference lies in the elements dimensions, which is likely to be connected with different design variables response during the optimal search. The maximum stress for the optimal pallet is 13.4 MPa, the discrepancy in rigidity criterion is 0.65 %. The values of the three first natural frequencies are presented in Table 2. As the table indicates, the lowest natural frequency of the pallet is nearly 9 times as high as the frequency of the constrained oscillations, which additionally proves the rightful limitation by the first natural frequency.

Table 1. The Results of Pallet Optimization

Pallet Design	Thickness, mm				Maximum vertical displacement, mm	Mass, tons
	Upper board	Side wall	Inside wall	Rib		
Serial	60.0	60.0	50.0	60.0	0.249	38.12
FEM	60.0	30.0	20.0	60.0	0.427	24.40
Original to be optimized	70.0	40.0	40.0	70.0	-	-
Optimal	29.0	36.3	36.3	69.5	0.452	24.59

Table 2. The Range of Optimal Pallet Natural Frequencies

The form of oscillations	Frequency, Hz		
	axis *x*	axis *y*	axis *z*
1	269.2	272.2	88.6
2	283.2	294.5	192.0
3	306.8	526.1	268.2

Sensitivity analysis [5]. To assess the impact of problem (2) limitations onto the design variables we investigated the behavior design variables variations around the optimal point decision. To do so we fix all variables but one and investigate the change in displacement, stress and frequency. The parameter change span is set to be ± 25 % in order to round the results for the practical use. The sequence of the design variables changes is given in Table 3. The change of limitation was defined in relation to the least value of the corresponding limitation while the design variable was changing from – 25 to + 25 %, i.e.

$$[(\psi_{+25\%} - \psi_{-25\%})/\ \psi_{\min}]100\ \%,$$

with

$$\psi_{\min} = \begin{cases} \psi_{+25\%},\ \text{if}\ \psi_{+25\%} < \psi_{-25\%} \\ \psi_{-25\%},\ \text{if}\ \psi_{+25\%} > \psi_{-25\%} \end{cases}$$

Table 3. The Results of Response Analysis

Design Variables	Response Limitation, %			
	Displacement	Stress	Stability	Frequency
The thickness of the side and inside walls (0.0363 m)	73.4	54.6	51.5	5.4
The thickness of the upper board (0.0290 m)	16.7	4.9	8.2	85.8
The rib thickness (0.0695 m)	15.2	10.4	25.1	0.2

The response analysis shows that if the pallet design was to be improved in terms of natural frequency, the best results could be achieved by the varying the upper board thickness. In other cases, best results are achieved by varying the thickness of side and inside walls. Using the response analysis information the designer can carry out a systematic analysis of the design and thus refine it. The received data supports the rigidity dominance over other criteria in the process of load-carrying structures design. It allows limiting the research efforts to one criterion, others being checked for performance consistency on the final stages of the optimal search.

b. Probabilistic Model

Deterministic mathematical models are widely used in calculation and design. However, spatial repositioning of a spot on the processed work piece, which is placed on the rotating table, depends on certain factors, namely on the center of gravity position in relation to the table pivot axis. Due to possible operational errors (e.g. work piece misplacement on the table or equipment misuse), the misalignment of the work piece center of gravity with the table pivot axis is of statistical character. The resulting eccentricity can be treated as a normally distributed variable with mathematical expectation $m = 0$.

The coordinates (x,y) of the point A (Figure 2) factual load application resulting from the work piece load form a system of random variables, normal density for which is expressed by the formula [6]:

$$f(x,y)=\frac{1}{2\pi\sigma_x\sigma_y\sqrt{1-r^2}}\times$$
$$\times\exp\left\{-\frac{1}{2\left(1-r^2\right)}\left[\frac{\left(x-m_x\right)^2}{\sigma_x^2}-\frac{2r\left(x-m_x\right)\left(y-m_y\right)}{\sigma_x\sigma_y}+\frac{\left(y-m_y\right)^2}{\sigma_y^2}\right]\right\} \quad (4)$$

Rectangular areas are characterized by elliptic eccentricity distribution, i.e. there is a dispersion ellipsis with a major a semiaxis and a minor b semiaxis. Assuming that semiaxes coincide with the coordinate axes, the origin of coordinates point C coincides with the dispersion center and random variables x,y are independent, the formula (4) reads:

$$f(x,y)=\frac{1}{2\pi\sigma_x\sigma_y}\exp\left[-\frac{x^2}{2\sigma_x^2}-\frac{y^2}{2\sigma_y^2}\right].$$

The full dispersion ellipsis is described by the equation [6]

$$\frac{x^2}{\left(4E_x\right)^2}+\frac{y^2}{\left(4E_y\right)^2}=1, \quad (5)$$

with $E_x \approx 0.675\sigma_x$, $E_y \approx 0.675\sigma_y$ being major probable deviations. For the variables normally distributed within the limits of $[-a;\ a]$ и $[-b;\ b]$ the values are $a/\sigma_x \approx 3$, $b/\sigma_y \approx 3$. Considering that $a = L/20$, $b = B/30$ the equation (5) takes on the following form:

$$\frac{x^2}{0.002L^2}+\frac{y^2}{0.0009B^2}=1.$$

In order to identify the point A coordinates we formulate the optimization problem:

to maximize F_i (6)

limited by $$\frac{x^2}{0.002L^2}+\frac{y^2}{0.0009B^2}=1.$$

The problem (6) is solved for the accepted dimensions L and B by Lagrange's method of multipliers [3], the resulting coordinates and eccentricity are:

$x = 0.21$ m, $y = 0.06$ m, $e = 0.218$ m.

The problem of the pallet optimal design in this case is stated similarly to the problem (2). Two occurrences are taken into account while calculating the probabilistic model of pallet loading:

1) the values t_c, t_p are limited by non-negativity, i.e. $t_c = t_p \geq 0$;
2) the values t_c, t_p are limited by the casting conditions in accordance with the formula [7]

$$t_{\min}=10\sqrt{\left(2L+B+H\right)/3}\text{, mm}, \quad (7)$$

with *L, B, H* being formation outer dimensions in meters. In our case: $t_c = t_p = t_{min} = 23$ mm.

Table 4 shows the calculation results of the pallet probabilistic model in comparison with the previously investigated models.

Table 4. The calculation results of the three models

Pallet Model	Thickness, mm				Rigidity	Mass,
	Upper board	Side wall	Inside wall	Rib	Discrepancy, %	tons
Serial	60.0	60.0	50.0	60.0	40.0	36.80
Deterministic	29.0	36.3	36.3	69.5	0.65	24.59
Probabilistic:						
$t \geq 0$	8.6	17.2	17.2	63.4	0.54	14.22
$t \geq 23$ mm	23.1	23.2	23.2	38.9	27.0	15.80

The acquired results demonstrate that in case the formation is exposed to ununiform load the probabilistic model calculation allows additional reduction of the formation mass without the loss of performance in comparison with deterministic model. The probabilistic model calculations require considering technological limitations, such as casting conditions, which determine the minimum wall thickness. The resulting formation rigidity reserve (27%) indicates the probability to search for additional designer arrangement improvement.

c. Considering the processed work piece rigidity

We have analyzed the deterministic and probabilistic approaches to the rotating table pallet calculation. While calculating models we used only those pallet load models, which considered solely the weight of the processed work piece, not its rigidity. However, the spatial repositioning of a spot on the processed work piece, which is placed on the rotating table, depends not only on the table rigidity, but also on the work piece rigidity.

Calculating the work piece rigidity assumes that the processed work piece is firmly fixed in pallet corner zones at three points (Figure 4, the application of force points), which provides for the fact that the two load schemes (considering and not considering the work piece rigidity) become equivalent.

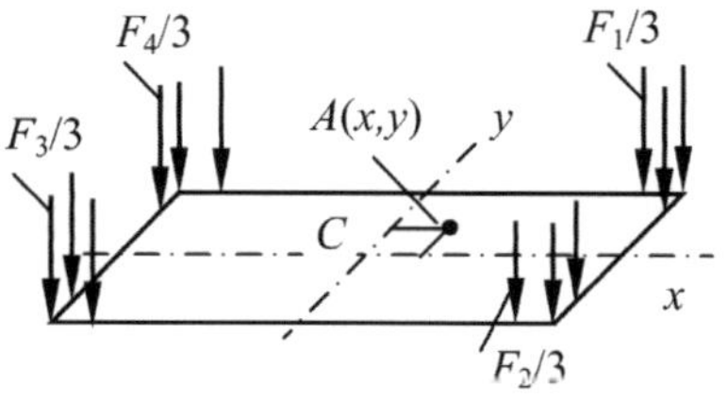

Figure 4. Calculation

To assess the combined action of the system 'pallet – work piece' we suggest using a conventional basic work piece of minimum rigidity (without bridges, ribs, internal closed contours etc.) with calculated weight of 2000 kN and the cross section providing for the specified center of gravity eccentricity *A* (*x,y*) with coordinates (Figure5):

$x = L/20 = 5.6/20 = 0.28$ m, $y = B/30 = 3.6/30 = 0.12$ m.

All the calculations took into account the limitation (7) on the walls and ribs thickness due to casting conditions. For the specified pallet the value is $t_{min} = 23$ mm.

The calculation is executed with the help of bundled software APM WinMachine (version 7.0) by the finite elements method:

1) for the pallet with optimal dimensions,
2) for the pallet with body wall thickness of 23 mm.

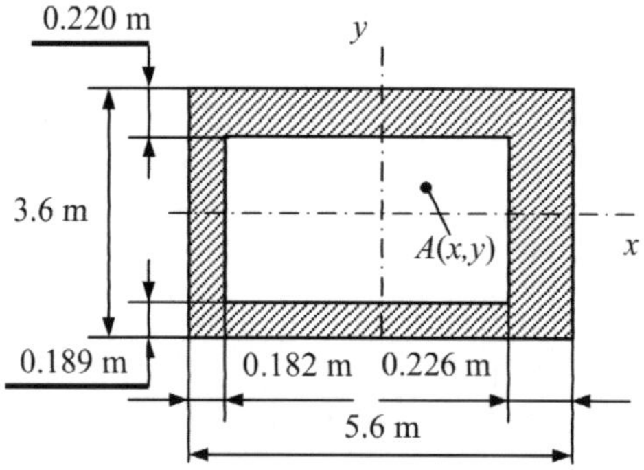

Figure 5. Cross-section of a conventional work piece

Figure 6 displays the pallet deformed state caused by the load of a conventional work piece. The calculation results given in Table 5 prove that the rigidity of the processed work piece considerably influence the rigidity of the pallet and, consequently, the load carrying formation of the table in general, making it possible to reduce pallet mass by 22.3 %. The minimum value of 23 mm thick pallet natural frequency is 88.18 Hz. It is significantly higher than permissible natural frequency of 10.8 Hz (defined by the spindle highest frequency 500 min^{-1} with 30% resonance offset). Therefore, when the pallet mass is reduced there is no possibility for resonance to appear during the machining.

The real processed work pieces do have various elements, which magnify their rigidity (e.g. walls, ribs, inside contours etc.), thus magnifying the rigidity of the 'pallet-work piece' system. However, due to the variety of possible work pieces arrangements, and, accordingly, various cross section rigidity it proves practical to calculate the design with the help of simplified conventional work piece with minimum rigidity for certain work piece nomenclature in order to obtain the rational table formation structure. The excessive rigidity of the real work pieces in comparison with the conventional work piece becomes the rigidity reserve of the table load-carrying system.

Figure 6. The deformed state of the pallet with a work piece

Table 5. Calculation results considering the processed work piece rigidity

Loading model	Pallet elements thickness: Upper board	Walls	Ribs	Pallet mass
	mm	mm	mm	tons
Not considering the processed work piece rigidity, optimal design	29.0	36.3	69.5	24.59
Considering the processed work piece rigidity	23.0	23.0	69.5	19.11

4. Conclusions

The analyzed calculation models of the heavy rotating table pallet demonstrate a wide range of rational design decisions in improving the load carrying systems of multi-purpose machines. Finite elements method in combination with optimization methods and considering the work piece rigidity on the grounds of a conventional basic work piece of minimum rigidity allow to design load-carrying machine structures without excessive properties.

References

[1] Atapin V G 2001 Proektirovanie nesushchikh konstruktsii tyazhelykh mnogotselevykh stankov s uchetom tochnosti, proizvoditel'nosti, massy [Structural design of heavy multi-task machines taking into account the accuracy, performance, weight]. *Vestnik mashinostroeniya – Russian Engineering Research*, 2001, no. 2, pp. **3–6.**

[2] Atapin V G 2001 Optimizatsiia nesushchei sistemy stola tyazhelogo mnogotselevogo stanka [Optimization of the support sistem of the heavy machining center table]. *Obrabotka metallov (tekhnologiia, oborudovanie, instrumenty) – Metal Working and Material Science*, 2006, no. 4 (33), pp. **30–32**.

[3] Ravindran A, Ragsdell K M and Reklaitis G V 2006 *Engineering Optimization: methods and applications* 2nd ed. New Jersey, John Wiley & Sons **688** p. ISBN-10: 0–471 –55814–1. ISBN-13: 978–0–471–55814–9.

[4] Bunday B D 1984 *Basic optimization methods*. London, Edward Arnold Publ. **136** p. ISBN-13: 978–0–713 –13506–0. ISBN: 0–713 –13506–9.

[5] Haug E J, Choi K K, Komkov V. *Design sensitivity analysis of structural systems*. Mathematics in Science and Engineering. Vol.177. Orlando, Florida, Academic Press, 1986. 381 p.

[6] Ventcel' E S 1999 *Teorija verojatnostej* (Theory of probabilities). Moscow, Vysshaja shkola, **576** p.

[7] *Metallorezhushie stanki* (Metal-cutting machine-tools) Moscow, Mashinostroenie 1986. **256** p.

IOP Conf. Series: Materials Science and Engineering **126** (2016) 012012 doi:10.1088/1757-899X/126/1/012012

Rational design of the column of a heavy multipurpose machining center

V V Atapin [*], N V Kurlaev

Faculty of Mechanical Engineering and Technologies, Novosibirsk State Technical University, 20 Prospekt K. Marksa, Novosibirsk, 630073, Russia

*E-mail: teormech@ngs.ru

Abstract. The main purpose in the design of supporting constructions of heavy multipurpose machining center is the reduction of mass at the given precision and productivity of machining. Accomplish these ends the technology of rational design of supporting constructions is offered. This technology is based on the decomposition method and the finite elements method in the combination with optimization methods. The technology has four stages: 1) calculation of external forces and loads, 2) as a result the boundary conditions (force, kinematics) for individual supporting constructions are formed, 3) a problem about final optimal distribution of a material by the individual supporting constructions with the real cross-section is solved; 4) dynamic analysis. By the example of design of the column of a heavy multipurpose machining center the main stages of rational design of the individual supporting constructions are shown. At a design stage of the carrying system consisting of load-bearing structures with simplified geometry, optimum overall dimensions of the column are identified. For the admitted system of preferences, it is necessary to accept the fact that the carrying system with the column with the sizes of cross section of 1.8 m (along x axis) and 2.6 m (along y axis) is the best. The analysis of the work of the column under the torsion condition with the use of method of mechanics shows that the column with square cross sections = 2.46·2.46 m which rigidity on torsion is 26 % higher in comparison with a production version is the best. The results of calculation show that a production-release design of the column with longitudinal and transverse edges of rigidity is 24 % heavier than the column with the edges located on a diagonally at equal rigidity.

1. Introduction

For the past decades the finite elements method has been widely used in the machine-tool industry for the static and dynamic analysis of supporting constructions of task geometry [1, 2]. Good results in the technology of rational design of supporting constructions of machine tools can be achieved when using the finite elements method together with the optimization methods [3–5]. The description and practical application of the widely used optimization methods are presented in [6, 7].

The overall dimensions of a metal-cutting machine tool are specified at the stage of project requirements. Then on the basis of calculations of strength and rigidity a designer determines sizes of supporting constructions that should meet the requirements for accuracy and productivity of machining. While analyzing configuration arrangements for individual supporting constructions general methods of solids mechanics, for example the method of strength of materials [8], are also effectively used.

In research paper [9] the technology of the design of supporting constructions of heavy multipurpose machining centers (Figure 1) is presented, it allows to design machines without excess capabilities, to predict their performance under the conditions of intense mechanical forces and to give scientific grounding to the choice of design options. The technology has four main stages:

1) determining external forces using deterministic or stochastic models;

2) calculating the set configuration of the carrying system of a machine taking into account the required productivity and accuracy of machining; supporting constructions here have simplified cross section geometry. As a result the boundary conditions (force, kinematics) for an individual supporting construction are formed, which makes it possible to analyse supporting constructions independently of one another;

3) the result of the stage of modeling an individual supporting construction is an optimal construction with real cross section geometry and with the minimum possible mass, provided the boundary conditions formed at the previous stage are satisfied;

4) dynamic analyzing or simulation modelling average operating conditions of the carrying system with optimal supporting constructions.

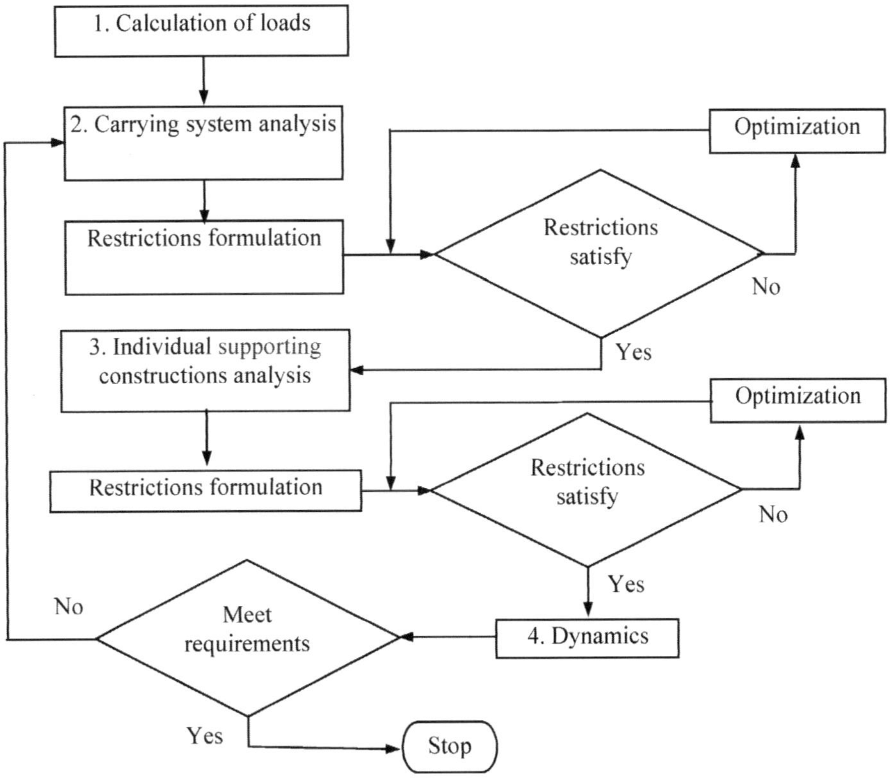

Figure1. The Design Technology of Machining Centers

The purpose of this paper is to calculate and analyse configuration arrangements for the most loaded supporting construction – the column of a heavy multipurpose machining center of a milling, drilling and boring group (Figure 2). To achieve the purpose the finite elements method, the optimization methods and the methods of strength of materials are used. A multipurpose machining center is a complex consisting of two items that are not put together with assembly operations at a manufacturing plant but designed to perform interrelated operating functions – a special milling-and-boring machine (Figure 2, Items 1-3) and a slewing-sliding table (Figure 2, Items 5-7). This fact makes it possible to do calculations for a table and a milling-and-boring machine separately.

2. Results and Discussion

a. Analysis of the response of supporting constructions as part of a carrying system of a milling-and-boring machine. The supporting constructions of this machine have a complicated configuration. These are space thin-walled constructions, which have differently directed stiffening ribs, a multiply connected closed outline, differences in wall thickness, and so on, that makes it hard to analyse directly their response to external actions as part of a carrying system. To make a comparative evaluation of the choice of external overall dimensions of the constructions their simplified cross-section geometry is presented. So, in particular, unlike the regular (production-release) construction (Figure 3), the column will have rectangular box-type cross-section without stiffening ribs and simply connected closed profile.

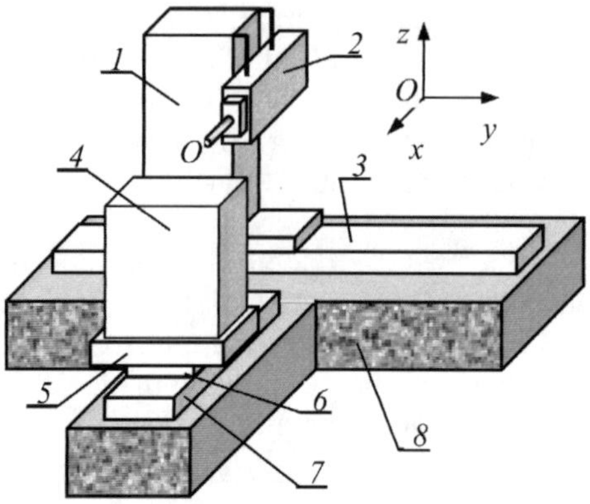

Figure 2. Machining center: *1* – column; *2* – spindle head; *3* – bed; *4* – work piece; *5* – pallet; *6* – slewing table.

Design conditions for carrying systems of machines are chosen based on the analysis of operating experience of machines that are similar in terms of configuration. Yet, the operations should be identified where accuracy and productivity are close to the limiting values. Relating to drilling, milling and boring machines the operation of this kind is face milling. A number of values for calculation are chosen as reference ones according to the project requirements for machine designing. In particular, the machine should provide the maximum forward force of 40 kN, for fine milling – 3 kN. The following design conditions are accepted for the carrying system of a machine:

- the spindle head is in the upper limit position provided the column is in the centre of the bed, extensions of the spindle ($\leq$ 0.4 м) and the slide ($\leq$ 0.6 м) correspond to the limiting values;
- external loading for a machine is cutting forces applied at point O on the spindle axis (Figure 2). The ratio for the components of cutting forces is $P_x : P_y : P_z = 0.5 : 1.0 : 0.7$;
- the bed rests on the resilient bearings (64 bearings), the rigidity of which is calculated the way that a bearing resists load of 40 kN, and the vertical displacement is $2\cdot10^{-5}$ m;
- contact deformations of moving joints is taken into account: column – spindle head, spindle head – slide, column – bed. The calculation of point O displacements (Figure 2) in the direction of x, y, and z axes is done for finish machining based on the rough surface rod model. As a result only due to contact deformations at the joints point O displacements of the spindle along x, y, and z axes are $\delta_x = 13.89\cdot10^{-6}$ m; $\delta_y = 8.11\cdot10^{-6}$ m; $\delta_z = 5.45\cdot10^{-6}$ m, respectively.

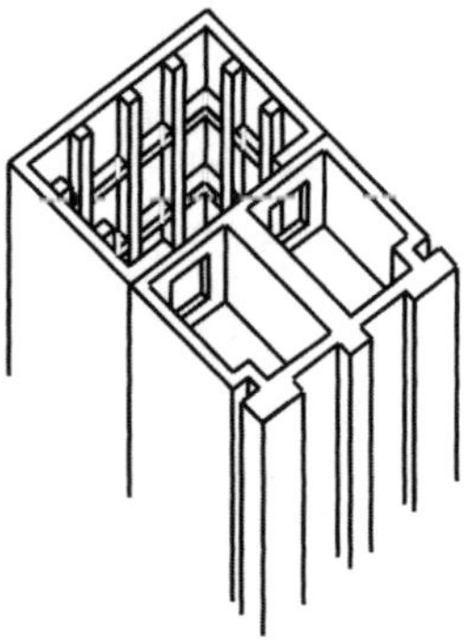
Figure3. The production version

The mathematical model of the carrying system of a milling-and-boring machine is formulated as a mathematical programming problem:

to optimize
$$f(X)=\sum_{i=1}^{n}\rho_i V_i \quad (1)$$

under constraints:

upon stress
$$g_1 = 1 - \sigma_{eqv} / \sigma_{allow} \geq 0, \quad (2)$$

upon point O displacements along axes

x $$g_2 = 1 - \Delta_x / [\Delta_x] \geq 0, \quad (3)$$

y $$g_3 = 1 - \Delta_y / [\Delta_y] \geq 0, \quad (4)$$

z $$g_4 = 1 - \Delta_z / [\Delta_z] \geq 0, \quad (5)$$

upon design parameters $$g_5 = V_i \geq 0, \quad i = 1, 2, \ldots, n, \quad (6)$$

where n – a number of plate-type finite elements; ρ – material density; V – the volume of the finite element; σ_{eqv}, σ_{allow} – equivalent stress and allowable stress; Δ_x, Δ_y, Δ_z, $[\Delta_x]$, $[\Delta_y]$, $[\Delta_z]$ – design and allowable (in brackets) point O displacements along x, y, z axes, respectively. The target function (1) is the construction mass. The design variable is thickness of the profile wall of supporting constructions.

When identifying allowable displacements $[\Delta_x]$, $[\Delta_y]$, $[\Delta_z]$ of the carrying system in the machining medium the departures from flatness and straightness on the machined surface are considered. In accordance with the accepted standards of accuracy when milling the tolerances for flatness and straightness are:

- on vertical displacement of the spindle head (5 m) – $60 \cdot 10^{-6}$ m;
- for the transverse displacement of the column (it is 4.5 m to the centre of the bed) – $60 \cdot 10^{-6}$ m.

According to the data obtained from the manufacturing plant the machining error at a cutting zone is distributed among a milling-and-boring machine and a table as 70 % and 30 %, respectively. When doing calculations it is impossible to consider all the factors that affect the machining accuracy, so, the safety factor for rigidity is taken as 1.5. Finally the tolerances for flatness and straightness for a milling-and-boring machine are $28 \cdot 10^{-6}$ m.

Point O displacements at a cutting zone are determined by the sum of displacements as a result of deformations of the carrying system itself and contact deformations at the joints. To calculate the carrying system point O displacements obtained as a result of contact deformations are subtracted from the obtained tolerance for point O displacements ($28 \cdot 10^{-6}$ m). The total is:

$$[\Delta_x] = 14.11 \cdot 10^{-6} \text{ m}, \ [\Delta_y] = 19.89 \cdot 10^{-6} \text{ m}, \ [\Delta_z] = 22.55 \cdot 10^{-6} \text{ m}.$$

While optimizing the carrying system of a machine it is believed that the displacements in the cutting zone along x, y, z axes must be less than or equal the corresponding allowable displacements. For the admitted system of preferences the option when the mass is the least and the displacements equal or near the allowable values is considered to be effective.

The problem of constrained optimization (1) – (6) is solved with the penalty function method as (r – penalty parameter)

$$\varphi(X, r) = f(X) + r \sum_{j=1}^{J} \left[1 / g_j(X) \right], \quad (7)$$

but using Davidon-Fletcher-Powell method to solve the problem of unconstrained optimization (7) [6, 7].

Figure 4 shows the deformed state of the carrying system of a milling-and-boring machine. To solve the problem with the finite elements method a plate-type four-node finite element is used. It can be seen that the column suffers the combined effect of bending and torsion and is the most loaded element of the carrying system of a machine. The data obtained while studying the carrying system utilizing supporting constructions of different type and for various operating conditions compared to the production versions are presented in Table 1. *The average operating conditions:* finish machining, face milling cutter 250 mm in diameter, the number of teeth is 20, the depth of cut is 0.5 mm, feed per tooth is 0.05 mm, spindle rotation frequency is 170 min^{-1}.

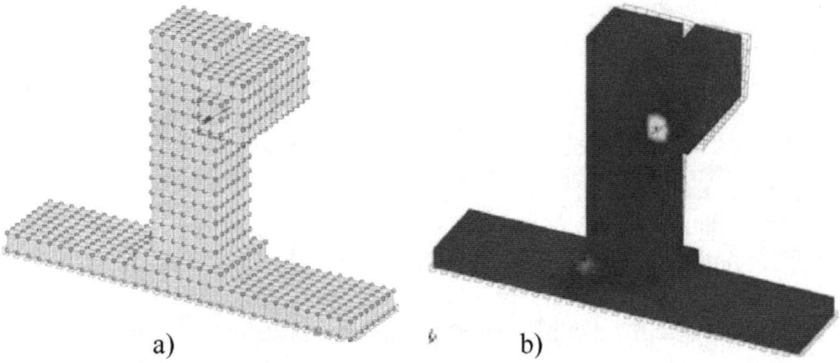

a) b)

Figure 4. Initial (a) and deform (b) states of the machining center.

The calculation results show that it is possible to improve the configuration of the supporting constructions compared to their production versions. For the admitted system of preferences it should be recognized that the carrying system with the column with the sizes of cross section of 1.8 m (along *x* axis) and 2.6 m (along *y* axis) is the best. The carrying system with the column of 1.6×2.8 (m) is rejected due to the great extent of the column section along *y* axis, as it brings to further decrease in the working stroke of the column along this axis. It should be stated that in the course of optimization point *O* displacement along *y* axis is an active constraint. If to take the average operating conditions as a basis while designing a machine, then when calculating the production version it is clear that the advantage by mass is about 30 %, with the standards of machining accuracy to be retained.

Table 1. Calculation Results for the Carrying System

Parameter	Operating Conditions					
	Limiting					Typical
Version	Production	Calculation				Production
Cross-section sizes (along axes), m:						
−column (*x*–*y*)						
−spindle head (*y*–*z*)	2.0·2.46	2.3·2.3	2.46·2.46	1.8·2.6	1.6·2.8	2.0·2.46
−slide (*y*–*z*)	1.3·2.20	1.8·1.9	1.30·2.20	1.7·2.0	1.7·2.0	1.3·2.20
	0.6·0.80	0.7·0.7	0.80·0.80	0.8·0.8	0.8·0.8	0.6·0.80
Point *O* displacements, micron:						
− *x* axis (norm 14.11)	7.34	6.75	6.43	9.51	9.58	6.28
− *y* axis (norm 19.89)	19.89	19.89	19.89	19.89	19.89	19.89
− *z* axis (norm 22.55)	11.58	11.74	11.56	12.10	12.69	11.20
Wall thickness, mm:						
− *yz, xz* plane	98	94	96	99	98	71
− *xy* plane	146	162	147	107	106.7	99
Mass, tons	169.9	179.7	181.0	158.2	157.2	119.3

b. Analysis of the configuration arrangements for the column

The production column (Figure 3) is a spatial thin-walled construction that consists of two parts connected with a bolted joint. The front of the column, which has guides for a spindle head, consists of two closed boundaries. The inner chamber of the back of the column has longitudinal and transverse edges crossing throughout the height and it is used to position a counterweight (it unloads vertical travel actuator of a spindle head from its weight).

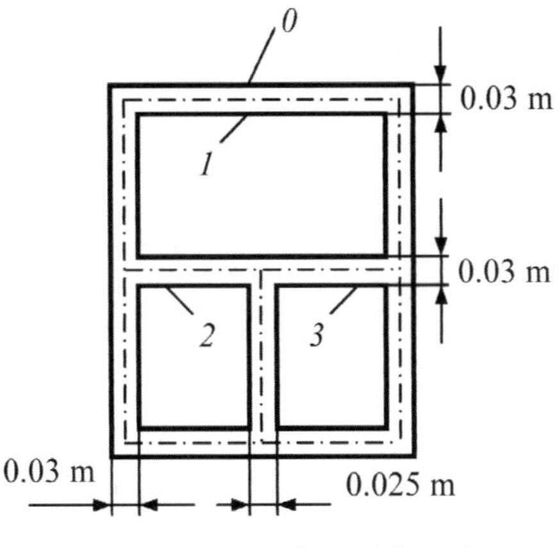

Figure 5. Cross section of the column

1. To test the multiply connected contour of the production column for torsional rigidity simplified representation of the column is presented (Figure 5) and Prandtl analogy is applied [8]. In the case of n–connected cross-section the torsional moment is calculated from formula

$$M = 2\left(\iint \varphi dxdy + \sum_{i=1}^{n-1} \varphi_i F_i - \varphi_0 F_0 \right), \tag{8}$$

where φ – Prandtl function; F_i, F_0 – areas bounded by inner and outer contours. In the case being considered formula (8) becomes as:

$$M = 2\sum_{i=1}^{n-1} f_i w_i = 2\sum_{i=1}^{n-1} f_i \varphi_i , \tag{9}$$

where f_i – area bounded by the contour, which everywhere halves the section thickness in cross-section; w_i – membrane slack surface function (w_i is identical with φ_i). The reference angle of torsion is calculated from formula

$$\theta = M/(GI_k), \tag{10}$$

where GI_k – cross-section rigidity for torsion. Figure 6 shows the calculation results (9) – (10) compared to the production version (Figure 6, a), rigidity for torsion of which is taken as 100 %.

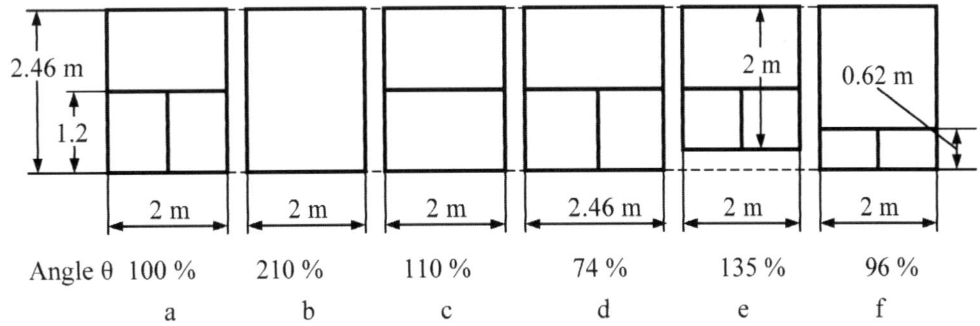

Figure 6. Cross sections of the column.

The data analysis shows that under these conditions the column with square cross section (Figure 6, d) which rigidity for torsion is 26 % higher in comparison with a production version is the best. In Fig.6, f there is the case where some improvement in rigidity also occurs. In reality the column with square cross section of 2×2 m (Figure 6, e) is also produced. However, section rigidity in this case is 35 % lower than rigidity for torsion of a production version.

2. The column can have various configurations of edges which to great extent determine its rigidity and, therefore, rigidity of a machine on the whole. A simplified construction of the column is considered to study the way a configuration of edges influences the column rigidity. Figure 7 shows the configurations of edges (there are no edges on the front wall). The configuration of edges in a production version is presented in Figure 7, a. The design model of the column simulates all kinds of deformation of a production version. Wall constraint upon displacements of the column structures for all configurations of edges is identical as the scheme of finite elements makes it possible to have a constant mesh for plate-like finite elements simulating column walls. The design data to do calculation: wall thickness is 0.04 m, edge thickness is 0.06 m, and column restraint along the lower contour is rigid.

If to denote, for instance, a turning angle of the front wall, which a spindle head slides along, by $\theta = |y_1 - y_2|/L_{12}$, where y_1, y_2 – linear displacements of point 1 and point 2, respectively, along y axis, L_{12} – distance between points 1 and 2 (Figure 7), M – column mass, the following results are obtained:

Configuration of Edges	θ, %	θ/*M*, %
Figure 7, a	100	100
Figure 7, b	70	70
Figure 7, c	100.3	124

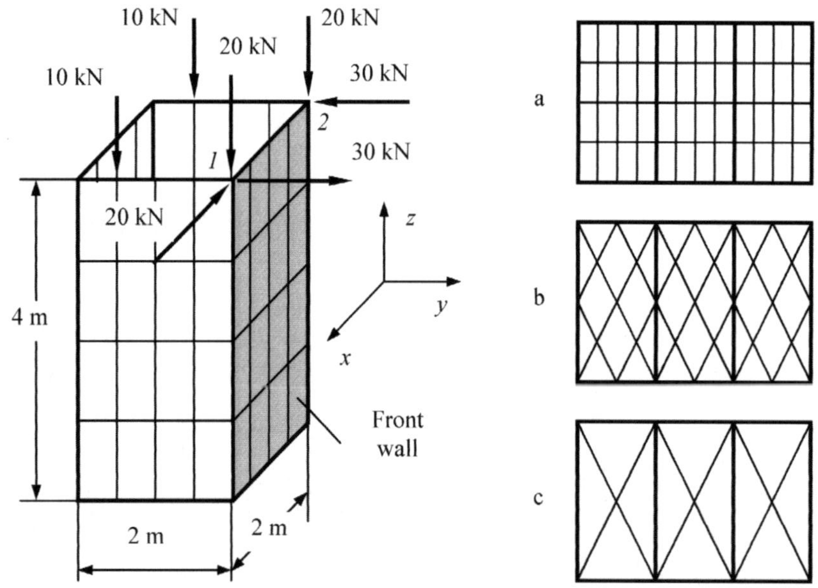

Figure 7. The design model of the column and the scan of the column chamber with different configurations of stiffening ribs

The results of calculation show that a production-release design of the column with longitudinal and transverse edges crossing throughout the height (Figure 7, a) is 24 % heavier than the column with edges crossing diagonally (Figure 7, c) at equal rigidity, and less rigid than the column with longitudinal and transverse edges crossing diagonally (Figure 7. b). Consequently, while designing a column slanted edges are far more preferable. As it is known, influence of longitudinal edges on shifts due to torsion is inappreciable, and due to bending it does not go beyond 10…20 %.

3. Conclusions

The results of calculation show that it is possible to improve configuration of supporting constructions, particularly a column, compared to a production version. For the admitted system of preferences, the carrying system with the column with the sizes of cross section of 1.8 m (along x axis) and 2.6 m (along y axis) is stated to be the best.

Studies of the impact of the configuration of edges of rigidity show that a production-release design of the column with longitudinal and transverse edges crossing throughout the height is 24 % heavier than the column with edges crossing diagonally at equal rigidity and less rigid than the column with longitudinal and transverse edges crossing diagonally. Consequently, while designing a column slanted edges are far more preferable.

References

[1] Lull B 1977 Statische und dynamische Berechnung von Werkzeugmaschinengestellen. *Maschinenbautechnik* **26(1)** 10-3.

[2] Roscher A 1978 Berechnung der dynamischen Eigenschaften von Werkzeugmaschinengestellen mit Hilfe der Methode der finiten Elemente *Maschinenbautechnik* **27(4)** 156–60.

[3] Rao S S and Grandhi R V 1983 Optimum design of radial drilling machine structure to satisfy static rigidity and natural frequency requirements *Journal of Mechanical Design* **105**, iss. 2 236–41. doi: 10.1115/1.3258515.

[4] Reddy C P and Rao S S 1978 Automated optimum design of machine-tool structures for static rigidity, natural frequencies and regenerative chatter stability *Journal of Manufacturing Science and Engineering* **100** iss. 2 137–46. doi: 10.1115/1.3439401.

[5] Yoshimura M, Takeuchi Y and Hitomi K 1984 Design optimization of machine-tool structures considering manufacturing cost, accuracy and productivity *Journal of Mechanical Design* **106**, iss. 4 531–7. doi: 10.1115/1.3258606.

[6] Ravindran A, Ragsdell K M and Reklaitis G V 2006 *Engineering Optimization: methods and applications.* 2nd ed. New Jersey, John Wiley & Sons **688** p. ISBN-10: 0–471 –55814–1. ISBN-13: 978–0–471–55814–9.

[7] Bunday B D 1984 *Basic optimization methods*. London, Edward Arnold Publ. **136** p. ISBN-13: 978–0–713 –13506–0. ISBN: 0–713 –13506–9.

[8] Atapin V G, Pel' A N and Temnikov A I 2006 *Soprotivlenie materialov* [Strength of materials]. Novosibirsk, NSTU Publ., **556** p. ISBN 5-7782-0605-4.

[9] Atapin V G 2001 Proektirovanie nesushchikh konstruktsii tyazhelykh mnogotselevykh stankov s uchetom tochnosti, proizvoditel'nosti, massy [Structural design of heavy multi-task machines taking into account the accuracy, performance, weight]. *Vestnik mashinostroeniya – Russian Engineering Research*, 2001, no. 2, pp. **3–6.**

Technology of Strengthening Steel Details by Surfacing Composite Coatings

V G Burov, A A Bataev, Kh M Rakhimyanov*, D O Mul
Faculty of Mechanical Engineering and Technologies, Novosibirsk State Technical University, 20 Prospekt K. Marksa, Novosibirsk, 630073, Russia

*E-mail: kharis51@mail.ru

Abstract. The article considers the problem of forming wear resistant meal ceramic coatings on steel surfaces using the results of our own investigations and the analysis of achievements made in the country and abroad. Increasing the wear resistance of surface layers of steel details is achieved by surfacing composite coatings with carbides or borides of metals as disperse particles in the strengthening phase. The use of surfacing on wearing machine details and mechanisms has a history of more than 100 years. But still engineering investigations in this field are being conducted up to now. The use of heating sources which provide a high density of power allows ensuring temperature and time conditions of surfacing under which composites with peculiar service and functional properties are formed. High concentration of energy in the zone of melt, which is created from powder mixtures and the hardened surface layer, allows producing the transition zone between the main material and surfaced coating. Surfacing by the electron beam directed from vacuum to the atmosphere is of considerable technological advantages. They give the possibility of strengthening surface layers of large-sized details by surfacing powder mixtures without their preliminary compacting. A modified layer of the main metal with ceramic particles distributed in it is created as a result of heating surfaced powders and the detail surface layer by the electron beam. Technology of surfacing allows using powders of refractory metals and graphite in the composition of powder mixtures. They interact with one another and form the particles of the hardening phase of the composition coating. The chemical composition of the main and surfaced materials is considered to be the main factor which determines the character of metallurgical processes in local zones of melt as well as the structure and properties of surfaced composition.

1. Introduction

Surfacing technology of hardening coatings which allow increasing the durability of steel details under intensive wearing is widely used in industry for more than 100 years. Nowadays a great variety of electrodes, wires and powder mixtures intended for gas thermal, electric arc and plasma coating of wear resistant materials on steel details are produced and offered on the market. The composition of surfacing materials as well as the composition of the main detail metal determines the structure and properties of the layer surfaced which presents the material as a steel matrix with evenly distributed hard particles of carbides, borides and intermetallics [1-4]. Hard particles determine the wear resistance of the surfaced layer. They are either present in the surfacing materials composition or formed from power mixture components as a result of chemical and thermal processes during surfacing. Compositions of surfacing materials and surfacing modes used determine the result obtained. Attempts to surface materials, the structure and properties of which were identical to those of sintered hard alloys with the framework of refractory ceramic particles in metallic bond are not successful up to now. First of all, it is explained by the fact that in surfacing the coating it is impossible to get the framework of refractory particles because the matrix material strengthened by phase particles, the hardness of which is in several times higher than that of the matrix, is always formed. Besides, the surfacing process is accomplished by forming the liquid phase, which is typical of metallurgical processes in micro volumes of surfaced and main materials. Technology of strengthening surface layers of steel details must solve the problem of compatibility between coating materials and the main metal in two aspects: thermo mechanical and physicochemical. The thermo mechanical aspect supposes the absence of expansion voltage in the coating formed and the presence of the transition zone between the coating and the main material, which excludes the composition destruction on the border – "the coating – the main material". The physicochemical aspect of compatibility between the material coating and the main metal means excluding the formation of fragile phases both in the coating and in the transition zone. Technology of strengthening steel details by surfacing the composite coatings provides their high service properties and is determined by the chemical composition of steel and the surfaced material.

2. Conditions of Experiments

2.1. Materials

Mixtures of refractory metal carbides with metallic powders as well as mixtures of metals forming carbides and borides with graphite, iron and other metals were used as surfacing materials. Table 1 presents the information on initial powder materials of which surfacing mixtures were produced.

Table 2.1. Chemical composition of powder materials (chemical elements and compounds) used in the experiments

Properties		Properties of Powder Material											
		WC*	WC [5,6]	TiC	B4C	B**	C***	Ti	Ta	V	Mo	Co	α-Fe
Particles size, μm	min	4	2	3	14	1	30	50	0.7	1	2	1	5
	max	13	5	13	28	30	80	90	5	30	5	2	30
Presence of Oxygen in particles, % per mass		<0.3	-	<0.3	-	-	<3.5	<0.7	<0.7	0.5	1.38	<1.4	< 4.7

Notes:
* content of fixed carbon 5.8 – 6.2 % per mass and free carbon – not more than 0.1 %;
** boron in the amorphous state;
*** graphite powder

Surfaced powder mixtures were laid on the surface of flat samples by different methods: by filling, through sprinkling the concentrated suspensions with further drying. To increase density, powder mixtures on the samples were subject to packing by the manual press before surfacing.

During surfacing heating in the air atmosphere salts MgF_2, CaF_2 and LiF were added to powder mixtures, which limited oxidizing processes in melt.

Steel with the carbon content of up to 0.8% without alloying components or containing not more than 1% of chrome was used as the main metal.

2.2. Methods of Surfacing

Surfacing of powder mixtures was conducted by different heating sources: in vacuum radiation furnaces, by the laser beam with a wavelength of 10.6 μm in the air and argon atmosphere, by the electron beam directed from vacuum to the air atmosphere.

Temperature and time modes of sintering were limited by appearing the liquid phase in the whole volume of powder composition laid on the sample of the main metal. The time period during which melt in the surface layer existed was minimized.

2.3. Assessment of Physical, Mechanical and Service Properties of Compositions

After visual assessment the hardness of surfaced coatings was measured under different loads, which allowed determining not only the ability of the coating to resist deformation under high specific loads but also served as an indicator of property variation in the coating volume – from the transition zone to the surface layer. Measurements were conducted on hardness meter 402 MVD on frontal and cross sectional cuts of compositions. The composition strength was estimated by device 8801, the destruction viscosity was determined on the pile driver Metro Com 06103300. To analyze cause-and-effect links between composition properties and surfacing modes the structure of surfaced layers was studied by the optical microscope Carl Zeiss Axio Observer Alm and scanning microscope Carl Zeiss EVO 50 XVP.

Wear resistance of surfaced coatings was determined by friction against fixed abrasive particles (silicon carbide and aluminum oxide) with the seed size of 80 … 100 μm by comparing with industrial sintered hard alloy and steel 40X after hardening with further low tempering.

3. Results and discussion

The main difference in methods of surfacing used consisted in temperature-and- time heating modes and cooling of steel samples with laid powder mixture. During the experiments it was discovered [7] that in surfacing powder mixtures containing tungsten monocarbide on steel detail compound carbides of type (Fe,W)6C which are very fragile are always present in the transition zone. The formation of these carbides takes place under temperatures of over 710°C. Despite their instability they are preserved in the surfaced composition after its cooling because of lack of carbon. Increase in the heating velocity and cooling by laser and electron beam heating allows reducing the size of complicated fragile carbides and providing good strength and high service properties of compositions [8, 9]. To exclude the formation of fragile carbides in the transition layers it is necessary to use the barrier layers between the coating and the main metal. Such layers must have a strong bond with the main metal and the surfaced coating. Saturation of the steel surface layer by boron is effective as the technology of forming barrier layers [10].

The most perspective technology of hardening surface layers of large-sized steel details is out-of-vacuum electron beam surfacing of coatings. As a result of it the hardening phase is created in the process of interacting initial surfaced powders with one another and with the main detail metal. The use of such technology allows reducing the duration of chemical and thermal modification processes of detail surface layers preserving the initial structure of the main metal. In surfacing powder mixtures of iron and amorphous boron [11 - 15], barrier layers of the necessary thickness, composition and structure are formed. Figure 1 shows the structure of the borated layer produced by surfacing the amorphous boron powder mixed with powder α-Fe. A high speed of electron beam surfacing of coatings on large-sized details allows excluding changes in the structure of the main metal. Borated layers with a thickness of 300 to 900 μm can be used both as barrier layers for further surfacing of composite coatings with hardening particles of tungsten carbide and as a wear resistant coating alone.

The investigations of electron beam surfacing conducted on medium-carbon steel 40X of wear resistant composite coatings from powder mixtures "Ti-C", "V-C", "Ta-C", "Ti-Ta-C", "Ti-Mo-C" allowed making a conclusion that the most perspective compositions are "Ti-Mo-C" and "V-C". Electron beam surfacing under the accelerating voltage of 1.4 MeV, the current of electron beam of 25 … 28 mA, the beam width of 50 mm and the scanning speed of the treated surface relative to the beam of 0.01 m/s allows forming coatings with a thickness of up to 2.6 mm. They have a matrix structure of the composite material hardened by martensite and particles of titanium and vanadium carbides which are evenly distributed and constitute up to 33% of the coating volume (Figure 2). The coating austenite matrix provides viscosity to the coating and keeps carbide particle firmly, preventing their breaking-off during tribotechnical influence.

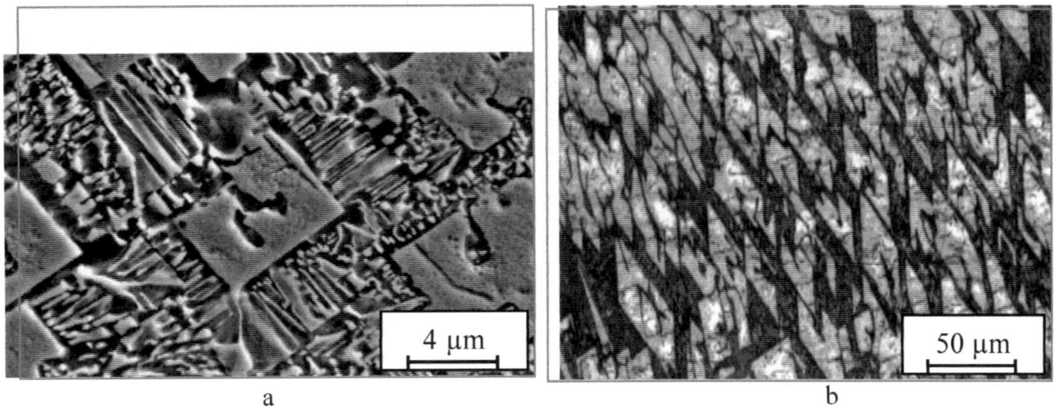

Figure 1. Structure of the steel 20 surface layer after out-of-vacuum electron beam surfacing of the powder mixture of amorphous boron and α-Fe: a – optical microscopy; b – electron scanning microscopy

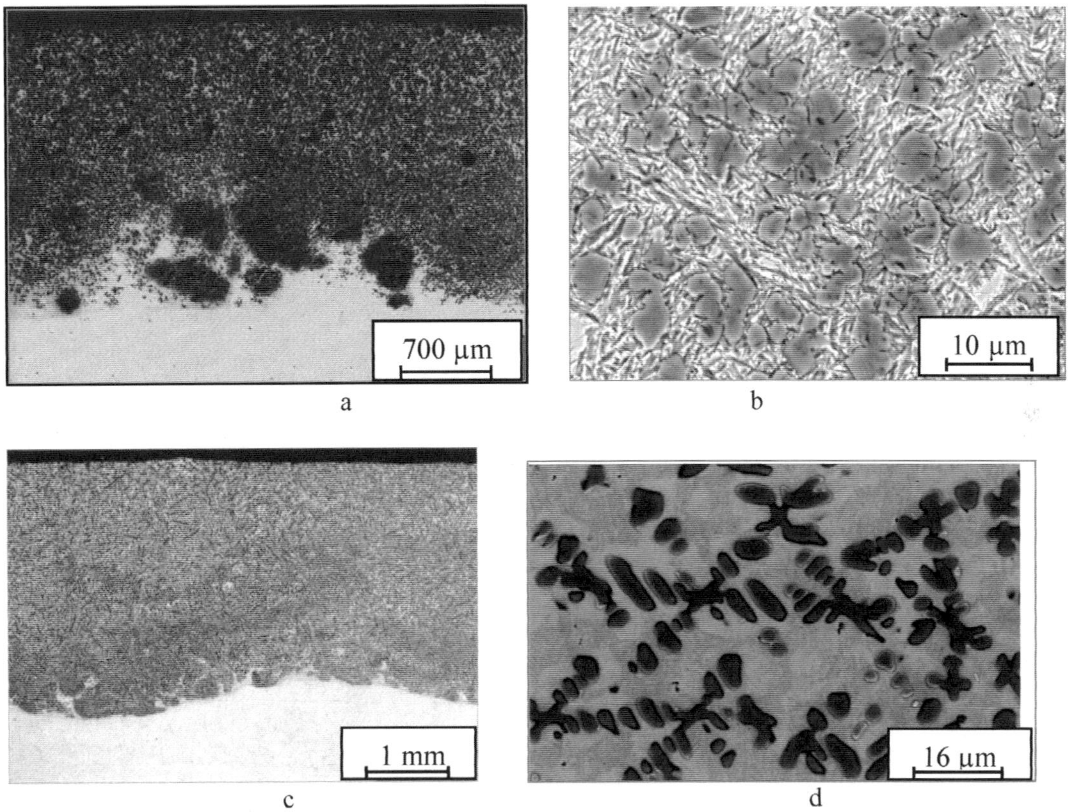

Figure 2. Structure of the steel 40X surface layer after vacuum electron beam surfacing: a, b – powder mixture Ti + Mo + C; c, d – V + C

The distinctive feature of surfaced coatings from the powder mixture "V – C" is the dendrite shape of vanadium carbide particles, while in surfacing the mixtures "Ti – Mo – C" carbides have a round shape similar to equiaxed. Under slipping friction the wear resistance of the coating with vanadium carbide particles is in 2.5 times higher than that of the material with round carbide particles of titanium carbide. Under friction against freely fixed abrasive particles the wear resistance of the material with carbides of the dendrite shape is three times higher than the wear resistance of the layer with equiaxed particles.

4. Conclusion

Technology of out-of-vacuum electron beam surfacing of composite coatings allows forming wear resistant layers with a thickness of more than 1…2 mm at the surfaces of large-sized steel details with high capacity. In this case there appears a possibility of using materials which during surfacing interact with one another, thus forming the particles of the hardening phase, evenly distributed along the volume of the surfaced layer.

References

[1] Kulkov S.N., Gnyusov S.F. 2006 Karbidostali Na Osnove Karbidov Titana i Volframa (Carbidosteels on the Basis of Carbides of Titanium and Tungsten) *(NTL,* Tomsk*)* p. 240

[2] Gnyusov S.F., Tarasov S.Yu. 2013 Structural Phase States and Heat Aging of Composite Electron Beam Clad Coatings *Surface and Coating Technologies* Vol. **232** 775-83

[3] Devoino O.G., Kardapolova M.A., Laber S., Feldstein E.A., D'yachenkova O.V. 2004 Wear Resistance of Coatings from Self-fluxing Alloys after Laser Doping under Dry Friction Conditions [Text] *Journal of Engineering Physics and Thermophysics* Vol. **77** Iss. **2** 372–6

[4] Wang D., Liang E., Chao M. 2008 Investigation on the Microstructure and Cracking Susceptibility of Laser-Clad V2O5/NiCrBSiC Alloy Coatings [J]. *Surface and Coatings Technology* **202** 1371–8

[5] Marusina V.I., Iskhakova G.A., , Rakhimyanov Kh.M. 1992 Phase and particle size composition of carbides formed during electric spark erosion machining of tungsten. *Soviet Powder Metallurgy and Metal Ceramics* Vol. **31** Iss. 10 870-73

[6] Iskhakova G.A., Marusina V.I., Rakhimyanov Kh.M. 1987 Determination of the microhardness of tungsten carbide particles produced in the spark discharge. *Soviet Powder Metallurgy and Metal Ceramics* Vol. **26** Iss. 10 852-4

[7] Burov V.G., Bataev I.A., Tyurin A.G., Veselov S.V. 2015 Structure and Properties of WC–Co Coatings Obtained on Steel Substrates by Liquid State Sintering in Vacuum [Text] *Surface Engineering* Vol. **31**, Iss. **7**. 540-4

[8] Gnyusov, S.F. Ignatov A.A., Durakov V.G., Tarasov S.Yu. 2012The effect of thermal cycling by electron-beam surfacing on structure and wear resistance of deposited M2 steel. *Applied Surface Science* Vol. **263** 215-22

[9] Gnyusov S.F., Durakov V.G., Ignatov A.A. 2013 Electron Beam Cladding by HSS R6M5 Powder *IEEE Transactions on Plasma Science* Vol. **41- 8**. 2196–200

[10] Burov V.G., Bataev V.A., Veselov S.V., Bataeva Z.B., Sameishcheva T.S. 2012 Surface Layer Formation by Melting Tungsten-Cobalt Powder Mixtures on Steel [Text] *Russian Engineering Research* Vol. **32- 1** 95–7

[11] Bataev I.A., Bataev A.A., Golkovsky M.G., Teplykh A.M., Burov V.G., Veselov S.V. 2012 Non-vacuum Electron Beam Boriding of Low-Carbon Steel [Text] *Surface & Coating Technology* Vol. **207** 245–53

[12] Krivezhenko D., Drobyaz E., Bataev I., Chuchkova L. 2015 Investigation of the structure and properties of boron-containing coatings obtained by electron-beam treatment *AIP Conference Proceedings* Vol. **1683** Art. 020104 (4 p.)

[13] Tyurin A., Nagavkin S., Malikov A., Orishich A. 2015 Microstructure of WC-Co hard alloy surface after laser treatment *Surface Engineering*. Vol. **31**, Iss. 1. 74-7

[14] Samoylenko V., Lenivtseva O., Polyakov I., Laptev I. 2015 Structure and mechanical properties of coatings fabricated by nonvacuum electron beam cladding of Ti-Ta-Zr powder mixtures *AIP Conference Proceedings* Vol. **1683** Art. 020198 (4 p.)

[15] Gromov V., Konovalov S., Bataev V., Ivanov Y., Kobzareva T., Semina O. 2015 The role of electro-explosion alloying with titanium diboride and treatment with pulsed electron beam in the surface modification of VT6 alloy *AIP Conference Proceedings* Vol. **1683** Art. 020093 (4 p.)

IOP Conf. Series: Materials Science and Engineering **126** (2016) 012014 doi:10.1088/1757-899X/126/1/012014

Inductor Hardening for Magnetic-Pulse Treatment of Tubular Parts

N V Kurlaeyv [*], K N Bobin, N A Ryngach, A. Kh. Rakhmyanov

Faculty of Mechanical Engineering and Technologies, Novosibirsk State Technical University, 20 Prospekt K. Marksa, Novosibirsk, 630073, Russia

[*] E-mail: kurlaev@corp.nstu.ru

Abstract. This paper focuses on the issues of modernization of standardized inductor construction for crimping tubular parts by the pulse electromagnetic field with the aim of increasing reliability of technique and its durability. There is given the description of the pilot model of the composite inductor for crimping tubular parts, as well as the results obtained during its test operation.

1. Introduction

High velocity forming methods that exist nowadays allow either to make more irregular parts for the less number of technological process steps or to make such parts that can't be obtained by any other methods of forming. [1]. Another advantage of forming by the pulse electromagnetic field is the quality of obtained parts that also positively influences the durability of metal treatment [2], as well as the bonding made by the method of crimping. Besides, pressure made by the pulse electromagnetic field has no direct influence on the surface of the objects under manipulation and, as a result, there exist no damages made during that manipulation unlike other kinds of high-speed forming. The disadvantages of this method include a necessity of high electrical conduction for work and inductor materials that restricts materials under manufacture by copper and aluminum alloys.

Techniques for manufacturing by the pulse electromagnetic field, i.e. inductors, experience great mechanic loading. Apart requirements for high electrical conduction, additional severe requirements are imposed to the durability of inductor materials. Reliability of monolithic-type inductors is due to reliability of current-conducting coil material. Beryllium bronze is mostly used as inductor material, as it possesses good electric conductivity and satisfactory mechanic features as compared to copper alloys. For instance, the most suitable coil material for inductors designated for making parts from alloy of Type D16AM with the dynamic yield stress $Y_0 \approx 200$ MPa is beryllium bronze with yield stress $\sigma_T = 500$MPa . There should also be considered that by electromagnetic treatment the pressure of the pulse electromagnetic field can outnumber the number of the dynamic yield stress Y_0 to 1.5...2.0 times. Herewith, bronze should be of primary smelting as the secondary bronze is of less durability because of included gas and porosity that come into the bronze during the smelting. Extra high tensile alloys of Type V95 if and others can't substitute bronze because of their poor longevity under impulse pressure. The making of current-conducting inductor elements from carbon and high-hardness steels is also possible; however, such inductors are of extremely low transmission efficiency when translating the greatest part of electrical power into the heat one.

The essence of the electromagnetic treatment is based on the Faraday Law of Electromagnetic induction. Energy during the electromagnetic treatment is stored in the capacitor bank, thus the current electricity flowing in short terms during its discharge. In its turn this current makes magnetic field which is compatible with the current guided in the workpiece. This interaction leads to the generation of Ampere force deforming the workpiece. Figure 1 shows an equivalent electrical circuit of the electromagnetic installment and presents a plain discharge resistor circuit where dying oscillation happens with the frequency of few tens of kilihertz.

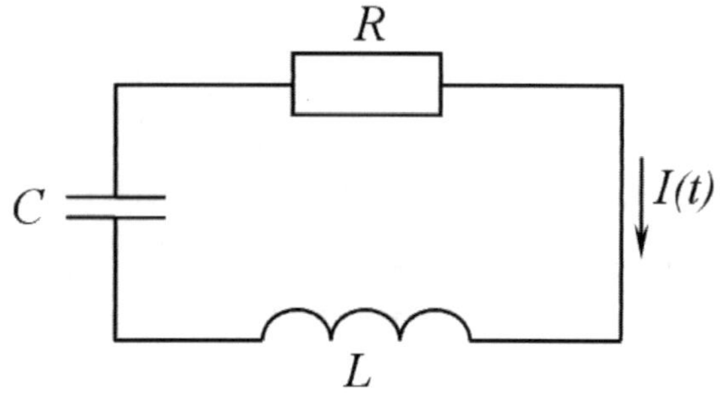

Figure 1. An Equivalent electrical circuit of the electromagnetic installment.

Time varying magnetic field with value B(0,t) penetrates into the metal of the workpiece, causing the eddy current, which current density i and resultant magnetic field depend on the depth x and time t (Figure 2). To describe the interaction of electromagnet with metal we will consider the current distribution through the skin depth.

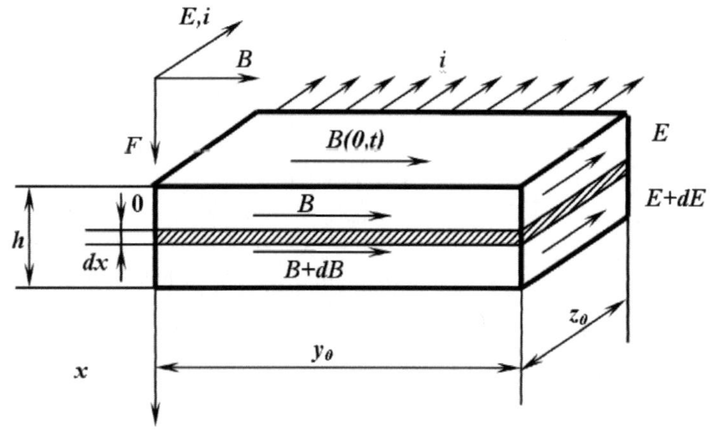

Figure 2. Interaction of nonstationary magnetic field with metal element.

For a layer with its depth dx Ampere and Faraday Laws can be formulated as follows:

$$i = -\frac{1}{\mu_0} \cdot \frac{\partial B}{\partial x} \tag{1}$$

$$\frac{\partial E}{\partial x} = -\frac{\partial B}{\partial t} \tag{2}$$

where μ_0 –is a magnetic constant equal to $1.26 \cdot 10^{-6}$ H/m, γ – is conductivity, a reciprocal to electrical resistivity, E – is electrical intensity , B – magnetic field strength.

From (2) taking into account the ratio i=γE, where γ = const differential equations of the interaction of electromagnetic wave with metal are received:

$$\frac{\partial B}{\partial t} = \frac{1}{\mu_0 \gamma} \cdot \frac{\partial^2 B}{\partial x^2} \tag{3}$$

Time varying magnetic field with value B(0,t) penetrates into the metal of the workpiece, causing the induced current, which current density i and resultant magnetic field depend on the depth x and time t:

$$B(x,t) = B_m \cdot \exp\left(-\frac{x}{\Delta}\right) \cdot \sin\left(\omega t - \frac{x}{\Delta}\right) \tag{5}$$

where Δ – is the depth of electromagnetic field penetration into the material, equal to the distance x, where magnetic induction amplitude and current density reduce e times; ω – rate of phase change of the electromagnetic field.

High frequency current through the inductor section at MIOM is distributed unequally [1]. When the frequency of changes in external field ω and metal conductivity γ increases electromagnetic field and current are concentrated in a thinner surface layer. The value of penetrating the electromagnetic field into the metal Δ is called a "skin-layer", and thus, it can be approximately considered that all currents coming through the inductor coil and workpiece material are concentrated in this layer of metal. Skin depth can be found with the help of expression (6).

$$\Delta = \sqrt{\frac{2}{\mu_0 \gamma \omega}} \tag{6}$$

According to the Ampere Law, force appearing in the unit volume of conductor is

$$\vec{f} = \left|\vec{i} \times \vec{B}\right| \tag{7}$$

The greatest intensity of the electromagnetic field can be defined when electromagnetic field pressure on the material is equal to the yield stress of inductor material.

$$B_T = \sqrt{2\mu_0 p_m} \tag{8}$$

when p_m – is material yield stress.

So, if the material of the inductor is copper M3, magnetic field strength is limited with volume 13.2 T. If it is an alloy of D16AT and beryllium bronze, then the created intensity of the electromagnetic field is limited with volumes 31.7 T and 22.4 T correspondently.

Operating frequency of current discharge of industrial installations for electromagnetic treatment is usually within the limits of 10 and 40 kHz, skin depth for copper and aluminum alloys under pulse electromagnetic field being less than 1 mm. Therefore, the weight of the inductor coil, that is mainly intended to guarantee the inductor and heat dissipation stiffness, has practically no influence on inductor electrical data of.

This paper focuses on one of the methods for increasing inductor durability at the expense of the composite construction which allows increasing operating pressure on the part under manipulation.

2. Results and Discussion

The construction of inductors for electromagnetic treatment is most commonly presented in the form of single- or multi-turn coils, made by mechanic work and either copper winding or copper alloys (bronze) one [4,5]. Besides, inductors are additionally covered with the special bondage that prevents from fragment dispersion in case of inductor emergency destruction. Figure 3 presents a typical construction of flat multi turn bronze. The given inductor is a multi-turn bronze coil of square section, embedded by polyurethane compound.

Figure 3. Overall view of a multi-turn inductor with a square-angled coil.

The division of the inductor coil into electric and power components allows increasing inductor stiffness and durability. In this case the electrical current flows through copper conducting loop, and the power impact of the electromagnetic field is accepted by the sound steel foundation. The main condition for effective work of such an inductor is a depth of copper layer, which must be at least 2 times thicker than a skin-layer at the operating frequencies of the electromagnetic installation.

The suggested conception can be realized with the help of a detachable single-turn coil inductor for crimping tubular parts (pencils of electric wiring harnesses) made according to OST 1 03967-81. The typical construction of a detachable single-turn coil inductor made from high-tensile-strength beryllium bronze OST 18175-78 became a basis for modernization. The inductor resistance of this construction is no more than 200 cycles, and then the destruction of the inductor happens. The main drawback of inductors of the suggested construction is rather high cost of beryllium bronze, as well as unrepairability of this construction – there appear thermal damages on the contacted surfaces while in service, that finally lead to the failure of the whole inductor. Inductor recycling or the recycled bronze usage reduces its resistance to 50-60 cycles.

Fig. 4 shows a modernized construction of a composite copper - steel inductor. Removable tapes from beryllium bronze play the role of the conductive part (3 and 4) 1.5 mm deep. The massive frame takes the power and thermal loading (1 and 2), which is made from steel alloy. The depth of conductive tapes is much higher than the skin depth at the operating frequencies of electromagnetic installment. Thus, the electrical current flows mainly through conductive tapes. The construction of the inductor allows changing malfunctioned copper tapes, whereas the amount of bronze used in the

IOP Conf. Series: Materials Science and Engineering **126** (2016) 012014 doi:10.1088/1757-899X/126/1/012014

inductor, is 15 times reduced as compared to the typical construction of the inductor for crimping pencils of electric wiring harnesses.

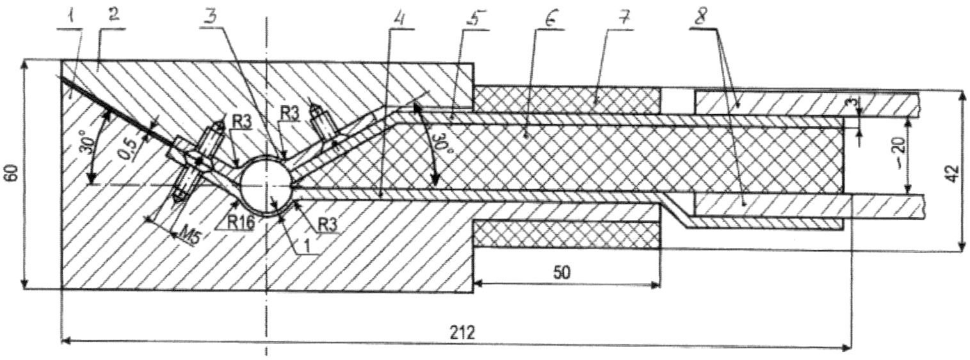

Figure 4. Diagram of a detachable single-turn coil inductor: 1 – half frame (30 HGSA steel); 2 - half frame (30 HGSA steel); 3 – current conductor of the operating zone (bronze BrB2); 4 – current conductor of the operating zone and и EMI (bronze BrB2); 5 – current conductor to EMI 6 – isolating distance bar (termazote); 7 – bandage (glass fiber cloth treated by epoxide compound); 8 - current conductors of EMI 20/5.

The trial run of the given inductor has revealed no substantial differences as compared to the regimes (power sored in the capacitor battery), got for the inductor totally made from bronze [1]. The construction of the inductor withstand more than 250 technological cycles without the breakage of the inductor steel frame.

To identify the durability of the composite inductors it is necessary to carry out some trials on inductor, however, the received results proved the workability and importance of the suggested conception. Besides, on the basis of this conception it is possible to improve even more the construction of inductors for electromagnetic treatment, as well as to create a composite multi-turn coil inductor for sheet treatment by the pulse electromagnetic field.

3. Conclusion

References The following conclusions were made on the basis of the trial run of the suggested composite inductor construction:

1. The construction of the composite inductor proved its workability.
2. While using the inductor of the suggested construction there is no necessity in changing the regimes of electromagnetic treatment – trials have not revealed any changes as compounds made according to the OST 1 03967-81.
3. Durability of composite inductor is 25% higher than durability of the inductor, made from beryllium bronze.
4. The suggested construction allows an easy change of damaged busbars that reduces the operation costs.
5. The content of the high-priced bronze in the construction is 15 times reduced as compared to the typical inductor construction for analogous purpose.

References

[1] Gulidov, A.; Ryngach, N.; Mishukov, A. 2001 Computer simulation of aircraft wires tips compression by pulse magnetic field *The 5-th Korea-Russia International Symposium on Science and Technology. -June 26-July 3, 2001,. Proceedings KORUS 2001 (Tomsk, Russia)* Vol. **1** 36-9

[2] Gonda V., Racz P., Horvath D., Mamalis A. 2014 Joining of tubular parts by electromagnetic forming: Computational investigations of strength, *8th Japanese-Mediterranean Workshop on Applied Electromagnetic Engineering for Magnetic, Superconducting, Multifunctional and Nano Materials, JAPMED 2013*, Vol. **792** 194-7

[3] J.H. Kim, D. Kim, M.-G. Lee, 2015 Experimental and numerical analysis of a rectangular helical coil actuator for electromagnetic bulging, *International Journal of Advanced Manufacturing Technology*, Vol. **78** 825-39.

[4] Siddiqui M.A., Correia J.P.M., Ahzi S., Belouettar S. 2008 A numerical model to simulate electromagnetic sheet metal forming process, *International Journal of Material Forming* Vol. **1** 1387-90

[5] Paese E., Geier M., Homrich R.P., Pacheco J.L., 2011 Simplified mathematical modeling for an electromagnetic forming system with flat spiral coil as actuator, *Journal of the Brazilian Society of Mechanical Sciences and Engineering*, Vol. 33, Iss. 3, 324-31.

IOP Conf. Series: Materials Science and Engineering **126** (2016) 012015 doi:10.1088/1757-899X/126/1/012015

Techniques for setting modes of thermal and deformation effect at combined hardening and finishing operations

Kh M Rakhimyanov[1*], **K Kh Rakhimyanov**[1], **A Kh Rakhimyanov**[1] and **A V Kutyshkin**[2]

[1] Faculty of Mechanical Engineering and Technologies, Novosibirsk State Technical University, 20 Prospekt K. Marksa, Novosibirsk, 630073, Russia
[2] Yugra State University

*E-mail: kharis51@mail.ru

Abstract. This paper considers the issues of setting the modes of thermal and deformation effects in the basic schemes at combined hardening and finishing operations. On the basis of solving the thermal physical problem of material high rate heating, the parameters of a thermo-hardened layer were determined within the range of the investigated modes. An algorithm for setting the mode parameters of high rate heating responsible for the hardening effect at the combined processing was proposed. The analysis of the mathematical model for forming a surface microrelief at ultrasonic deformation showed that the sizes, the form of fragments and the density of a microrelief were determined by the processing kinematic parameters. An algorithm for setting the rotation speed and feeding at ultrasonic deformation according to microrelief characteristics was developed. The conditions to form a completely regular microrelief on the processed surface that represent the ratio between a single imprint diameter at the ultrasonic deformation and the processing kinematic parameters were determined. The complex of the algorithms suggested for setting the mode parameters of high rate heating and ultrasonic deformation constitutes the techniques for setting the modes of combined hardening and finishing operations.

1. Introduction

Recent years has much contributed to the development of the effective methods for materials processing based on combining the high rate heating and plastic deformation [1, 2]. High energy techniques for creating a surface layer with high structural strength are effectively applied in technological processes of manufacturing the parts intended for using under severe operating conditions.

The techniques for combining thermal deformation processes during the surface processing of metals and alloys is given in the work [3], in which the principles and properties for developing new processing methods are described, and the criteria of combining high energy processes are stated. That allowed the development of basic schemes for combined processing which determines the location and the time of a thermal and deformation source operation in the zone of processing. Thus, the basic scheme I (TD) presupposes the deformation (D) effect from the thermal (T) source within the thermal cycle. In basic schemes II (D+T) and III (T+D), the introduction of the succeeding source of the combination into the zone of processing is performed only after all the processes induced by the preceding source in the surface layer material have been completed.

Due to reducing the number of various schemes which combine thermal and deformation sources to the three basic ones, the theoretical and experimental studies of any processing appear to be possible with the use of an integrated structure of complex modeling, which is given in detail in the work [3].

The development of foundation for theoretical research in the combinable thermo- deformational processes in the form of a complex mathematical model [1, 3] allowed the authors to determine the specific features of a development mechanism of any of the processes, identify the level of their interference, discover the emerging limitations and offer the ways of their overcoming. As a result, the developed theoretical foundation can serve as a basis for predicting the results of the combined processing of metals and alloys of various classes so as to ensure the required level of structural strength, stress state and surface microgeometry. The paper is devoted to the development of techniques for setting the thermal and deformational sources modes in the studied schemes of combined hardening and finishing of metallic materials.

2. Results and Discussion

The important stage of preparing the technological process related to setting the necessary processing modes is to obtain the generalized parameters of the technology under study which show the range of the results achieved together with the estimation of possible expenditures of energy and processing rate.

Taking into account that the thermal source used for the combined hardening and finishing operations is of surface type, the power distribution over the heating spot can be described as follows:

$$q(r) = q_0 \cdot e^{-kr^2} \tag{1}$$

The radius of the heating spot ($r_н$) is taken as the value of the radius (r), on which the power density q is equal to 0.05 q_0, where q_0 is the maximum power density in the centre of the heating spot. Then the source concentration coefficient (k), characterizing the form of the normal distribution curve is $k \cong 3/r_н^2$.

The kinetics of heating and cooling of a material surface layer in conditions of high rate heating by the concentrated power sources should be considered from the point of the classical theory of thermal conductivity. In this case the equation of heat flux distribution in a differential form appears as:

$$VC\frac{\delta T}{\delta z} = \frac{\delta}{\delta x}\left(\lambda\frac{\delta T}{\delta x}\right) + \frac{\delta}{\delta y}\left(\lambda\frac{\delta T}{\delta y}\right) + \frac{\delta}{\delta z}\left(\lambda\frac{\delta T}{\delta z}\right);$$

$$\lambda\frac{\delta T}{\delta x}\Big|_{x=0} = 0; \qquad \lambda\frac{\delta T}{\delta x} - \beta(T - T_0)\Big|_{x=X} - 0;$$

$$\lambda\frac{\delta T}{\delta y} - \beta(T - T_0)\Big|_{y=0} = 0; \qquad \lambda\frac{\delta T}{\delta y}\Big|_{y=Y} = q; \tag{2}$$

$$T\big|_{z=0} = T_0; \qquad \lambda\frac{\delta T}{\delta z}\Big|_{z=0} = 0; \qquad \lambda\frac{\delta T}{\delta z}\Big|_{z=Z} = 0,$$

where T = T(x, y, z) – the temperature of the part; T_0 – the ambient temperature; V – the thermal source path speed; q = q(x, z) – the density of heat flux rate on the outer surface of the part; λ = λ(T) – heat conductivity of the material being processed; C = C(T) – the volumetric specific heat of a material; β – the heat transfer coefficient; X, Y, Z – the sizes of a computational domain.

In order to calculate the required temperature distribution, the numerical method for solving the boundary value problem has been utilized (2). The results obtained due to solving the thermo physical problem of a material heating by the concentrated energy source have made it possible to determine the interrelation between high rate heating parameters and such characteristics of the process as specific expenditures of energy and processing rate at achieving the maximum thickness of a thermo-hardened layer at the modes excluding the melt formation on the surface of the material (Figure 1).

IOP Conf. Series: Materials Science and Engineering **126** (2016) 012015 doi:10.1088/1757-899X/126/1/012015

These results illustrate the technological capacities of the thermal source at different values of the concentration coefficient for achieving the maximum thickness of a thermo-hardened layer. In addition, the expenditures of energy and processing rate have been estimated.

The calculation of the thermophysical problem allowed finding the maximum achievable values of the thermo-hardened layer thickness within some ranges of processing modes (the concentration coefficient and the processing rate). This is shown graphically in the upper part of Figure 2.

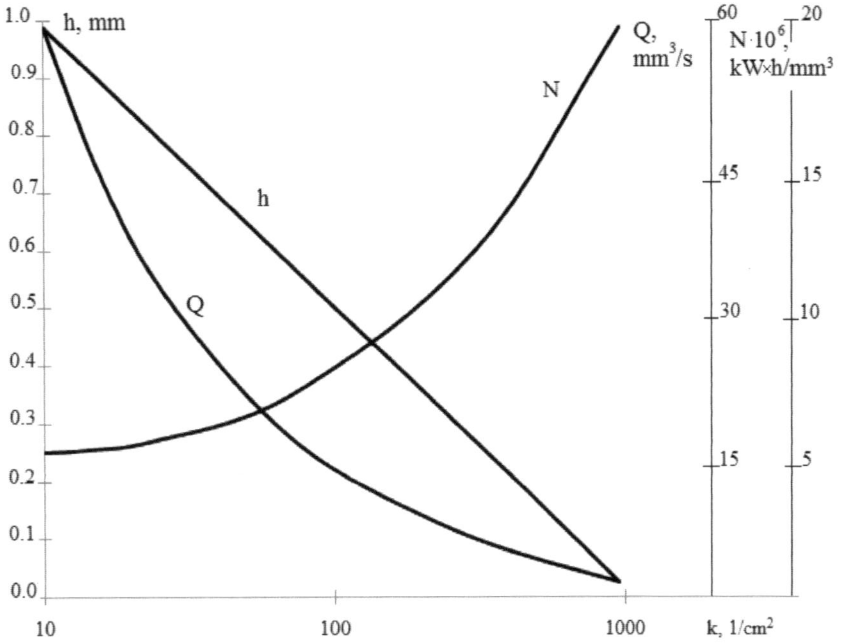

Figure 1. The ratio between the maximum thickness (h) of the thermo-hardened layer of steel 45, processing rate (Q), specific energy consumption (N) at the effect of the heating source with a different concentration coefficient (k) at the processing rate V = 1 cm/s.

As seen from Figure 2, within the chosen range of the investigated high rate heating modes, the thermo- hardened layer with the thickness of about 1mm is formed that is considered sufficient to operate the parts and tools in the conditions of friction and abrasive wear. The calculations show that, for all variants of processing which ensures the formation of the thermo-hardened layer of the uniform thickness, the power density of the thermal source remains unchanged. It is the ratio between the source concentration coefficient and the processing rate that changes.

The analysis of the results obtained by calculating the thermophysical problem concerning the estimation of the width of a thermo-hardened zone (d_{hard}) on the surface indicated its dependence only on the source concentration coefficient (see the bottom of Figure 2).

Thus, the results obtained by calculating the thermophysical problem and presented in Figure 3 can be applied for setting technological modes of high rate heating according to the preset values of the thermo-hardened zone width and the layer thickness. In this case, the algorithm for setting the modes is as follows: $d_{hard} \rightarrow k \rightarrow h \rightarrow q_0 \rightarrow V$ presented in Figure 2 schematically by arrows. So, the source concentration coefficient is determined by the value of width of the thermo-hardened zone on the surface of the part specified by technical requirements, and the power density of heating and the processing rate is determined by the required thickness of the thermo-hardened layer.

It should be noted that the surface of a part can significantly exceed the sizes of the thermo-hardened zone covered by the power concentration source.

In this case the multiple-pass processing with the overlapping of thermo-hardening zones should be set.

The deformation effect of the ultrasonic spherical deformer in the combined processing schemes is aimed both at creating the hardening effect and at forming the surface microgeometry while processing. The choice of ultrasonic vibrations of the spherical deformer as a source of deformation effect in the schemes of the combined processing is accounted for its technological advantages in achieving the hardening and finishing effect in comparison with other methods of surface plastic deformation [4-8].

The mathematical description of the microgeometry formation process at ultrasonic plastic deformation of parts (performed according to the turning scheme) is based on the model of imposing the plastic imprints (d_{imp})of a spherical deformer on each other when the kinematic motions – rotation speed (V_d) and feeding (S) are applied to the part. As a result, a microrelief consisting of separate fragments (ABCDEF) is formed (Figure 3).

The number of fragments per surface unit (microrelief pattern density) is defined as:

$$N = \frac{60 \cdot f}{V_d \cdot S} \quad (3)$$

where f – the frequency of ultrasonic vibrations, Hz; V_d – the rotation speed of a part , mm/s; S – the feeding, mm/rev.

The height characteristic of surface roughness (h_R), given as a surface roughness parameter (R_z) takes the form:

$$h_R = R_z \cdot 10^{-3} = \frac{D_S}{2} - \sqrt{\frac{D_S^2 - l^2}{4}} \quad (4)$$

where D_S – the diameter of the ultrasonic spherical deformer, mm; R_z – surface roughness parameter, mkm; l – a step of the microrelief in the direction of V_d, mm.

Taking into account that $l = \frac{V_d}{60 \cdot f}$ and transforming the expression (4), we get:

$$V_d = 120 \cdot f \sqrt{\frac{D_S^2}{2} - \left(\frac{D_S}{2} - R_Z \cdot 10^{-3}\right)^2} \quad (5)$$

The expression (5) identifies the speed characteristics (V_d) as one of processing kinematic parameters which determines the conditions for achieving the required roughness.

The value of the second kinematic parameter – the feeding (S), on which the value of microrelief pattern density depends (N), is determined from the expression (3):

$$S = \frac{60 \cdot f}{V_d \cdot N} \quad (6)$$

Figure 4 shows the scheme of setting the kinematic parameters of ultrasonic deformation according to the specified characteristics of microgeometry (the height parameter and density of the microrelief pattern). Thus, the bottom of Figure 4 presents the dependence of the characteristic curve of the roughness parameter (R_z) on the processing rate (V_d) at various values of the diameter of the ultrasonic spherical deformer. The upper part of Figure 4 shows the relationship between V_d and S, which ensure the creating of the microrelief with a certain pattern density.

The algorithm for setting the kinematic modes of ultrasonic deformation shown by arrows in Figure 4 is the following: $R_z \rightarrow V_{dmax} \rightarrow V_d \rightarrow N \rightarrow S$.

It must be aware of the fact that the value V_d determined according to the expression (5) is the maximum permissible value for ensuring the required surface roughness.

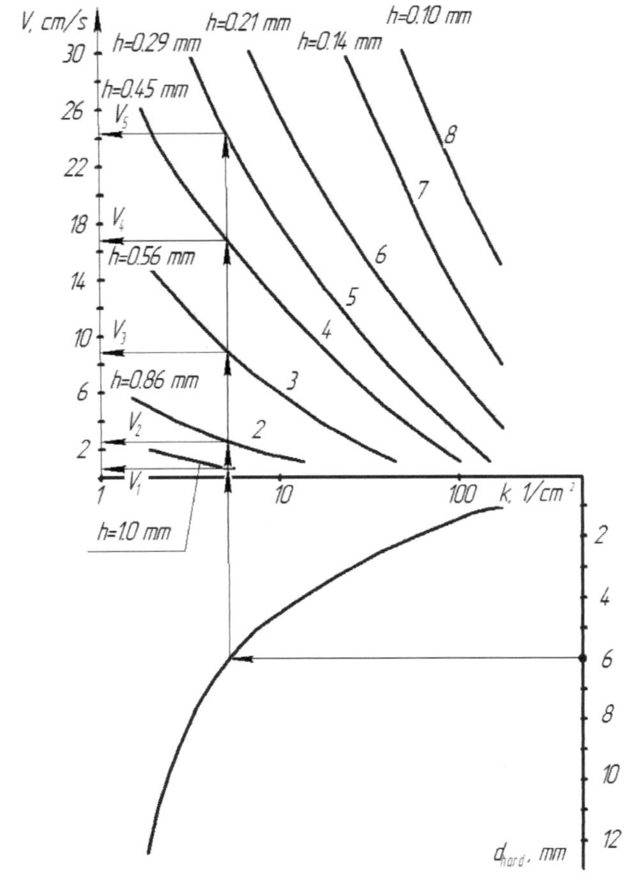

Figure 2. The correlation of parameters of the thermo-hardened layer of carbon steels on the coefficient of the source concentration and the modes of heating, the speed of treatment (V, cm/s) and the power density (q_0, W/cm^2): 1 – 3500;
2 – 4300; 3 – 7500; 4 – 10000; 5 – 13500; 6 – 15500; 7 – 17500; 8 – 20000.

The technology for ultrasonic hardening and finishing operation implies the formation of an entirely regular microrelief on the processed surface without any residual traces of previous processing.

The condition for forming this microgeometry is the limitation:

$$d_{imp} \geq \sqrt{\frac{S^2 + l^2(1-m)}{\cos\beta}}, \tag{7}$$

where m – the fractional part of the microrelief fragment which closes the unit turn of processing traces on the cylindrical surface. The value m = 0 – 1.0 determines the form of the microrelief fragment [9]; β – the angle of line inclination connecting the centers of the first and the last fragments in the neighboring turns of processing traces to the long axis of the cylindrical surface (Figure 3). The value of the angle is defined as:

$$\cos\beta = \frac{S}{\sqrt{S^2 + m^2 \cdot l^2}} \tag{8}$$

After transformation the expression (7) takes the following form:

$$d_{imp} \geq \sqrt{\left\{S^2 + \left[\frac{V_d \cdot (1-m)}{60 \cdot f}\right]^2\right\}\sqrt{1+\left(\frac{m \cdot V_d}{60 \cdot f \cdot S}\right)^2}} \quad (9)$$

The right part of the expression (9) presents a set of kinematic processing parameters interrelated by the characteristics of the microrelief to be formed. The value of unit imprint (d_{imp}) being intended with a spherical deformer can be calculated as:

$$d_{imp} = 2 \cdot \sqrt{D_S \cdot h_{max}} \quad (10)$$

where h_{max} – the maximum depth of the spherical deformer penetration into the processed surface; it depends on deformation mode parameters of ultrasonic processing of materials with different original hardness.

The results of modeling the ultrasonic deformation process are presented in works [10].

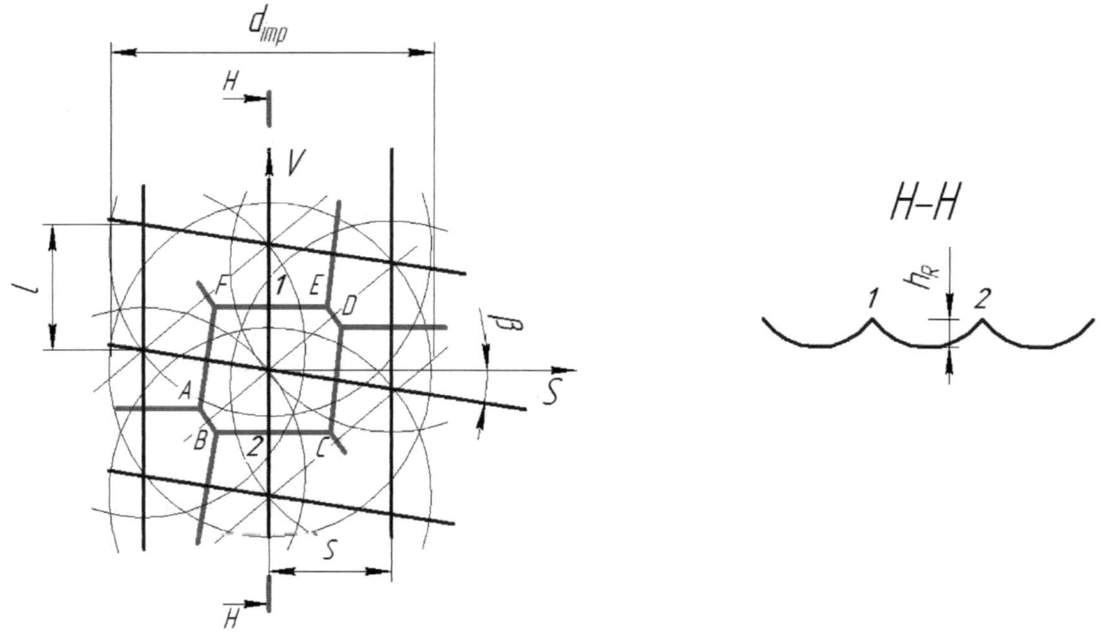

Figure 3. The scheme of microrelief fragment formation on the surface treated by ultrasonic deformation.

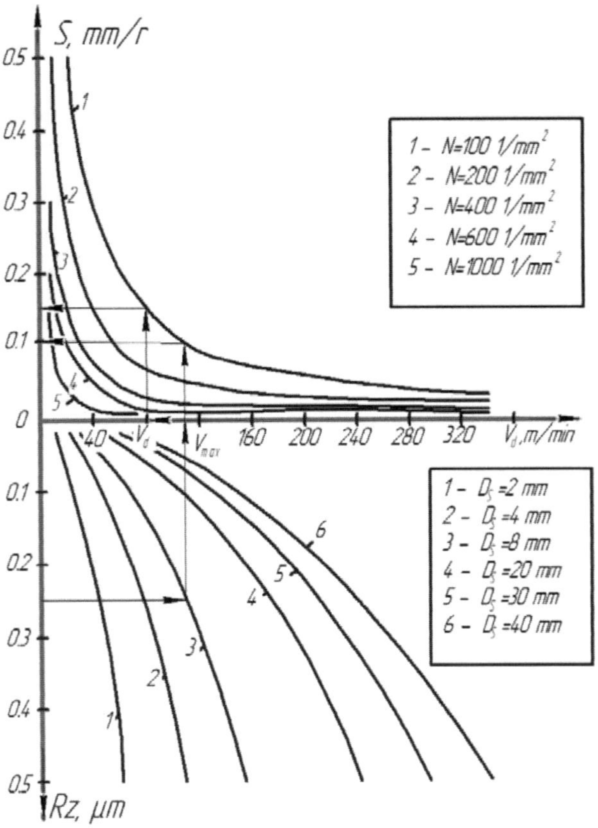

Figure 4. The scheme for setting the kinematic modes of ultrasonic deformation according to the microgeometry characteristics (f = 20 kHz).

3. Conclusion

The algorithms presented for determining the parameters of thermal and deformation processes to achieve the required hardening and finishing effects can be considered as techniques for setting mode parameters of the combined thermal deformational processing according to various basic schemes. In this case, the quantitative values of thermal source operating parameters both the power and operating ones irrespective of their basic scheme of treatment will be determined by the part specification according to the parameters of the thermo-hardening zone.

The values of kinematic parameters of ultrasonic deformation independent of the implemented basic scheme of the combined processing are determined only by the surface microgeometry requirements. The deformation modes of the ultrasonic processing specifying the necessary diameter of the plastic imprint depend on the hardness of the material deformed, which differs significantly in the basic schemes of the combined thermo-deformational processing.

References

[1] Liu Y, Wang D, Deng C, Huo L, Wang L and Fang R 2015 Novel method to fabricate Ti-Al intermetallic compound coating on Ti-6Al-4V alloy by combined ultrasonic impact treatment and electrospark deposition *Journal of Alloys and Compounds* **628** 208-12

[2] Rakhimyanov Kh M, Iskhakova G A and Rakhimyanov A Kh 1996 The role of ultrasonic plastic deformation in spark alloying *Fizika I Khimiya Obrabotki Materialov* **1** 68-72

[3] Solonenko O P, Marusin V V, Alkhimov A P, Orishich A M, Rakhimyanov Kh M, Salimov R A, Shchukin V G and Kosarev V F 2000 *High Energy Processes of Materials Treatment* **18** (Novosibirsk: Nauka, SIF RAN)

[4] Mordyuk B N, Prokopenko G I, Milman Y, Iefimov M O and Sameljuk A V 2013 Enhanced fatigue durability of Al-6Mg alloy by applying ultrasonic impact peening: Effects of surface hardening and reinforcement with AlCuFe quasicrystalline particles *Materials Science and Engineering A* **563** 138-46

[5] Yin D, Wang D, Jing H and Huo L 2010 The effects of ultrasonic peening treatment on the ultra-long life fatigue behavior of welded joints *Materials and Design* **31**(7) 3299-307

[6] Huo L, Wang D, Zhang Y and Chen J 2000 Investigation on improving fatigue properties of welded joints by ultrasonic peening method *Key Engineering Materials* **187(2)** 1315-20

[7] Mordyuk B N and Prokopenko G I 2007 Ultrasonic impact peening for the surface properties management *Journal of Sound and Vibration* **308(3-5)** 855-66

[8] Wang T, Wang D, Huo L and Zhang Y 2006 Subsection method of fatigue design for welded joints treated by ultrasonic peening *China Welding* **15(2)** 25-30

[9] Blyumenshtein V Y, Zaides S A and Rakhimyanov Kh M 2007 *Tekhnologicheskie protsessy poverkhnostogo plasticheskogo deformirovaniya* (Irkutsk: IrGTU)

[10] Rusinko A 2011 Analytical description of ultrasonic hardening and softening, *Ultrasonics* **51(6)** 709-14

Hybrid Processing: the Impact of Mechanical and Surface Thermal Treatment Integration onto the Machine Parts Quality

V Yu Skeeba[1], V V Ivancivsky[1], A V Kutyshkin[2] and K A Parts[1]
[1] Faculty of Mechanical Engineering and Technologies, Novosibirsk State Technical University, 20 Prospekt K. Marksa, Novosibirsk, 630073, Russia
[2] Institute of Management Systems and Information Technology, Yugra State University , 16 Chekhova str., Khanty-Mansiysk, 628012, Russia

E-mail: skeeba_vadim@mail.ru

Abstract. The comparative analysis of the two hybrid process technologies, which are based on the integration of mechanical treatment (abrasive grinding or turning) and a surface heat strengthening by high frequency current on the same processing equipment, is given in the paper. The acquired results demonstrate that the suggested integrating approach allows carrying out the processing on the one technological base, which leads to the increase in the quality of the machine parts surface layer. The conducted experimental research proves that a minor stock allowance value for the final mechanical processing (sparking out or diamond smoothing) ensures the absence of defects such as local abatement zones and provides strain hardening of the work piece surface. This leads to the formation of the work-hardened layer of 0.01 - 0.03 mm, increase in microhardness value by 12 - 17% and the level of residual compressive stress in the surface layer by 10 - 21 % respectively.

1. Introduction

The most promising direction of modern metalworking equipment development is to combine on a single machine multiple processing steps, and even processes of various types and sequences [1 – 7]. This idea implements a number of combined processing methods: electromechanical, friction, laser and ultrasound, plasma-ultrasonic, hardening grinding and others [8 – 15]. The use of combined treatment allows achieving some additional effect that is not characteristic of each of the merged operations separately.

The main characteristics of the parts surface layer quality are known to be formed on the final stage of their manufacturing process [16 – 19]. Therefore, this paper considers two realization schemes of integrated processing on the same production equipment. The first scheme is implemented on the basis of a lathe: rough turning, surface hardening by high frequency currents, finish turning and diamond smoothing. The second scheme is implemented on the basis of grinding machine: grinding, surface hardening, sparking.

The aim of the research is to study the impact of the integrated mechanical and surface thermal treatments operations on the hardness and residual stress of the steel machine parts surface layer.

2. Materials and methods

Plates with dimensions 100x10x5 mm and cylindrical bars with diameter d = 50 mm and length l = 70 mm made of 45 grade steel were used as specimens. The composition of the base material was defined on the optical emission spectrometer ARL 3460. The analysis results are given in Table 1.

The retrofit of 3G71 surface-grinding machine and UT16PM screw-cutting lathe by equipping the machine systems with additional concentrated energy source was carried out to integrate the processes of mechanical and surface thermal treatment. As the energy source we used a detachable hardening contour providing high-energy heating by high frequency currents (HEH HFC) powered by HFG 6 - 60/0.44 generator with current operating frequency $\omega = 440 \cdot 10^3$ Hz.

Table 1. Chemical compositions of 45 grade steel

Elements	C	Si	Mn	S	P	Cr	Ni	Cu
Weight, [%]	0.44	0.23	0.61	0.013	0.019	0.11	0.15	0.17

In the process of surface heat strengthening we used the loop inductor equipped with grade N87 ferrite (Figure 1). The heating was executed in accordance with deep-lying scheme by continuous step-by-step method. The thickness of the hardened layer did not exceed the depth of current penetration into the heated metal – 0.6…0.8 mm. The machining conditions are: the source power density $q_s = (1.5 - 4.0) \cdot 10^8$ W m^{-2}; the speed of the specimen movement under the inductor influence $V_s = (0.05…0.1)$ m s^{-1}. The breadth of the active inductor cord was $B_s = 2$ mm, the processing was executed with clearance $\Delta = 0.1…0.2$ mm. The intensive water (shower) surface cooling was used [20, 21].

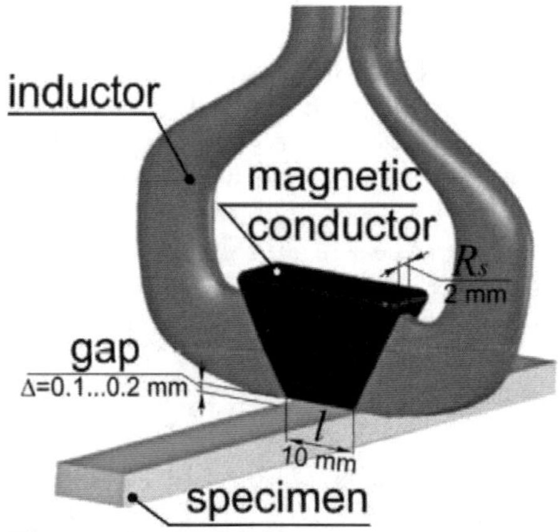

Figure 1. Loop inductor with ferrite magnetic core.

Abrasive grinding of the hardened and nonhardened steel was carried out taking into account the recommendations in [22] by the abrasive wheel in accordance with the State Standard "GOST P 52781-2007: 1 250x32x76 25A F60 L 6 V 35 m/s 2cl.; 1 250x32x76 25A F46 L 6 V 35 m/s 2cl.". The machining conditions are: wheel speed (cutting speed) of $V_{cs} = 35$ m/s, line feed was varied in the range of $V_{lf} = 5$ ... 20 m/min. Since the height of the wheel is greater than the breadth of the workpiece (10 mm) the scheme used plunge grinding. In this case, the cross-feed S_{cf} was not used, and the infeed grinding at the depth t was performed discretely per each table stroke.

Rough turning was performed by a straight cutting tool with indexable-insert, the material of the insert being T15K6, at the following modes: the cutting speed $V_{cs} \approx 92$ m min^{-1} ($n_d = 588$ min^{-1}); feed per revolution $S_{rev} = 0.35$ mm; the cutting depth t = 1 mm.

At the stage of finish turning we used a straight cutting tool with indexable-insert, the material of the insert being Al_2O_3-TiC. The cutting modes were: the cutting speed $V_{cs} \approx 133$ m min^{-1} (n_d = 882

min^{-1}); feed per revolution S_{rev} = 0.025 mm; the cutting depth t = 0.01÷0.015 mm. Sulfurized mineral oil *Sulfofrezol* was used as lubricant and coolant at the stages of rough and finish turning.

At the stage of diamond smoothing we used a toolholder with elastic nose with a settled diamond nosepiece manufactured in accordance with Standard Specifications "TU2-037-631-88". The sphere radius was R = 1 mm. Taking into account the rigidity of the lathe, surface layer rigidity of the processed workpiece after the surface hardening by HEH HFC (HV = 700…800) and the diamond sphere radius the radial component of the smoothing force P_y was 50, 100, 150, 200 N correspondingly. The circumferential speed of the workpiece was V_{cs} = 25.3; 33.2; 41.9 m min^{-1} (n_d = 168; 220; 278 min^{-1}); feed values were S_{rev} = 0.018…0.08 mm. Industrial oil of *I - 20A* grade was used as a lubricant and coolant during the process of diamond smoothing.

Structural investigation was conducted on the optic microscope *Carl Zeiss Axio Observer Z1m* and scanning electron microscope *Carl Zeiss EVO 50 XVP,* which was equipped by a energy dispersive detector *INCA X-ACT (Oxford Instruments).* To reveal the specimen microstructure we used a 5% ethanol solution of nitric acid and a saturated solution of picric acid in ethanol with surfactants [23].

To assess microhardness of the specimen hardened surface layer we used the instrument *Wolpert Group 402MVD*. A mechanic destructive method (layerwise electrolytic specimen etching) was employed to investigate the residual stress as well as X-ray technique, executed on the high definition diffractometer *ARL X`TRA* [24, 25]. The surface layer defects detection was performed on each technological stage by: visual optical method using microscope *Carl ZeissAxio Observer A1m;* capillary method; eddy-current testing employing eddy-current flaw detector VD-70. Surface topography assessment was executed on the laser profilograph-profilometer *Zygo New View* 7300. The surface deviation in form, undulation and asperity was measured by *Taylor Hobson* profilograph-profilometer *Form Talysurf Series 2.*

3. Results and discussion

The assignment of rational mode for each stage of the mentioned hybrid treatment was based on the recommendations in papers [26, 27] and conditioned by the maximum production capacity and the desired quality of the final product. For example, while choosing the second stage modes - surface hardening by HEH HFC limited by the required hardening depth h = 0,6 mm and a rational character of residual stress distribution - the power density (q_s [W m^{-2}],) and the source traverse speed (V_s [m s^{-1}]) were defined via a set of equations solution, namely functional dependence of the hardened layer depth on the HEHHFC modes $h(q_s, V_s)$ and functional dependence of the transition layer relative value on the surface hardening modes $\Psi(q_s, V_s)$ [28]:

$$\begin{cases} h(q_s,V_s) = a + bV_s + cq_s + dV_s^2 + eq_s^2 + fV_s q_s + gV_s^3 + hq_s^3 + iV_s q_s^2 + jV_s^2 q_s, \\ \Psi(q_s,V_s) = k + lV_s + mq_s + nV_s^2 + oq_s^2 + pV_s q_s + rV_s^3 + sq_s^3 + tV_s q_s^2 + uV_s^2 q_s, \\ 0{,}25 \le \Psi(q_s,V_s) \le 0{,}35. \end{cases}$$

with a = 0.426008, b = 2.827121, c = $3.025072 \cdot 10^{-9}$, d = -301.591960, e = $-4.694423 \cdot 10^{-18}$, f = $3.600666 \cdot 10^{-8}$, g = 1953.668810, h = $3.216427 \cdot 10^{-27}$, i = $1.375401 \cdot 10^{-17}$, j = $-3.779403 \cdot 10^{-7}$, k = 0.087564, l = -7.429933, m = $1.062284 \cdot 10^{-8}$, n = 235.19293, o = $-3.424286 \cdot 10^{-17}$, p = $-8.850919 \cdot 10^{-8}$, r = -1309.3045, s = $2.9423 \cdot 10^{-26}$, t = $1.403793 \cdot 10^{-16}$, u = $1.010925 \cdot 10^{-7}$ - being the values of functional dependency ratios for the 45 grade steel. The relative breadth of the transition layer Ψ equals the ratio of the transition zone (i.e. the zone between pure martensite structure layer and a layer with the original structure) to the depth of the hardened layer.

In the process of hardening the grade 45 steel at the depth of 0.6 mm the range of the recommended modes is limited by the source power density intervals $q_s = (3.0…3.4) \cdot 10^8$ W m^{-2} and speed V_s = (0.072…0.081) m s^{-1}. The calculated combination of the treatment modes ensures the required results in the hardening depth and a rational transition zone value.

The machines were adjusted for the following modes: 1) for the surface grinding machine 3G71 - $q_s = 3.2 \cdot 10^8$ W m^{-2} и $V_s = 0.078$ m s^{-1}; 2) for the screw-cutting lathe UT16PM - $q_s = 3.2 \cdot 10^8$ W m^{-2} и $n_d \approx 31$ min^{-1} (the reequipment of the main motion drive by adding the *HF Inverter model F*1500-G0015*S2B* frequency converted to the electric circuit was executed to provide the required rotation speed of the spindle).

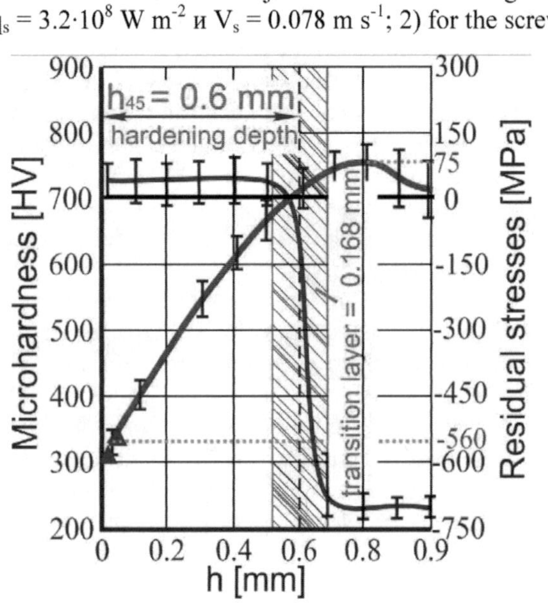

Figure 2. The distribution of microhardness and residual stress in the specimen surface layer after the hardening by HEHHFC.

Figure 2 illustrates the pattern of microhardness and the surface layer residual stress distribution in the specimen. The maximum level of compressive residual stress on the surface was σ = -560±20 MPa. The peak of the tensile stress was situated at the depth of ~ 0.8 mm and comprised σ = 75±45 MPa.

Figure 3a. shows the hardened grade 45 steel layer microstructure. It is made of martensite with insignificant additions of minor ferritic grains of less than 10 μm. The average microhardness level of the hardened by HEH HFC layer is 739 HV in comparison with initial material microhardness of 204 HV. The microstructure of the transition zone near the base material consists of ferritic grains of 20-30 μm and low-carbon steel (Figure 3b). The breadth of this zone is ~ 0.17 mm. The presence of ferrite and its lineage structure (the same as in the base material) indicates that this zone was heated between Ac_1-Ac_3.

The integral processing on 3G71 machine presupposes the removal of the stock allowance occurred due to volume expansion of the hardened layer (not more than 0.015 mm) to be performed at finish grinding. Considering the types of the used abrasive disks, the average abrasive grain rounded radius is: for the wheel - 1 250x32x76 25A F60 L 6 V 35 - 19 ± 4 μm; for the wheel - 1 250x32x76 25A F46 L 6 V 35 - 32 ± 5 μm.

Hence, the stock allowance for the finish grinding is significantly smaller than the abrasive grain average radius, which in fact excludes the possibility of mechanical metal removal and predetermines its plastic deformation.

Figure 3 The microstructure of the grade 45 steel after HEHHFC surface hardening:
a) hardened layer area; b) transition zone area.

The next stage was sparking, which was performed during τ = 15…20 s at the following modes: wheel speed (cutting speed) V_{cs} = 35 v s^{-1}, line feed V_{lf} = 20 m min^{-1}, cross feed S_{cf} = 0.2 mm per double pass. In the course of experimenting, we registered the increase in surface microhardness to the level of HV 830±15 (Figure 4), i.e. the increase was by ~ 12%. At the same time, there was an increase in the surface compressive residual stresses: σ = -620±10 MPa. The increase in the level of the residual stresses in comparison with the stage of HEH HFC hardening was by ~ 10%. The surface undulation was about Ra = 0.15±0.05 μm (Figure 5).

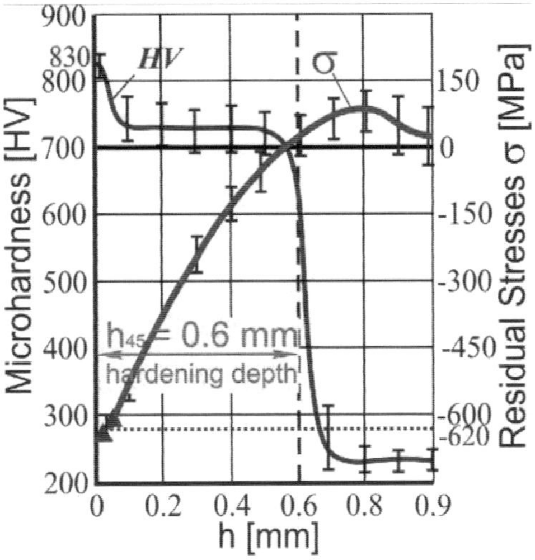

Figure 4. The distribution of microhardness and residual stresses in the surface layer after sparking τ = 15 s.

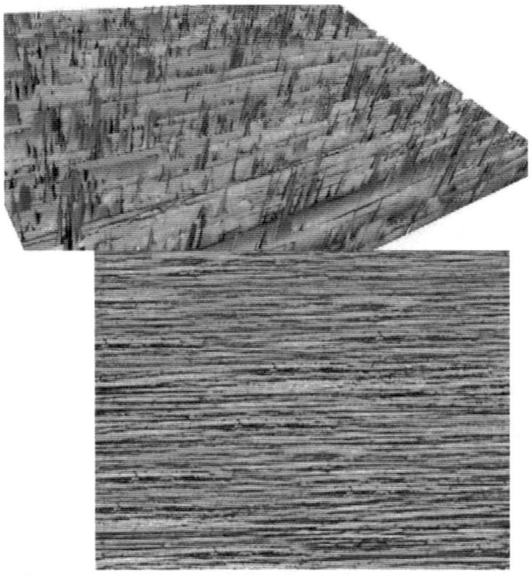

Figure 5. 3D model of surface topography after the stage of sparking. Surface undulation Ra = 0.143 μm. Sparking time τ = 15 s.

In the course of conducting experiments on work station based on UT16PM lathe the finish turning was performed to the size of the workpiece d = 48 mm, the surface undulation in Ra value was 1.2±0.2 μm. The surface microhardness and residual stresses were at the level of the HEHHFC stage.

In the process of diamond smoothing there forms a surface with plastic deformation and a rounded microrelief without tears and pits (Figure 6).

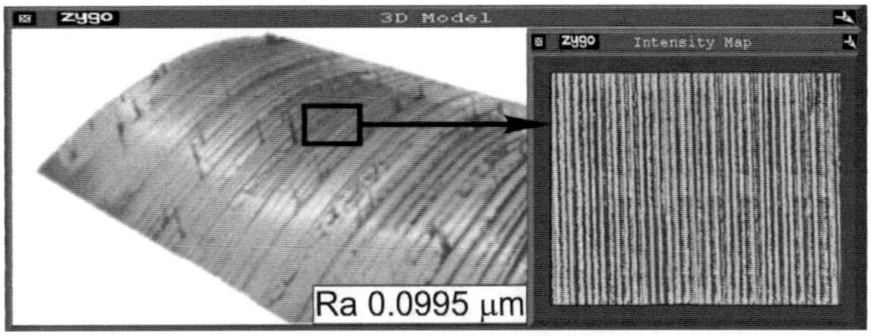

Figure 6 3D model of workpiece surface topography after the diamond smoothing stage: the initial surface undulation before the diamond smoothing Ra = 0.18±0.08 μm. Processing modes: cutting speed V_{cs} = 25.3 m/min; feed S_{rev} = 0.018 mm; smoothing force P_y = 150 H; R = 1 mm.

The structural research revealed that in the process of diamond smoothing there forms a work-hardened layer of 0.01…0.03 mm in the surface hardened specimen (Figure 7). The empirical data shows that the surface microhardness of the workpiece rises to 868 HV (the increase by~ 17%), and the level of compressive stresses in the surface layer rises to σ = -678±20 MPa (the increase by 21%).

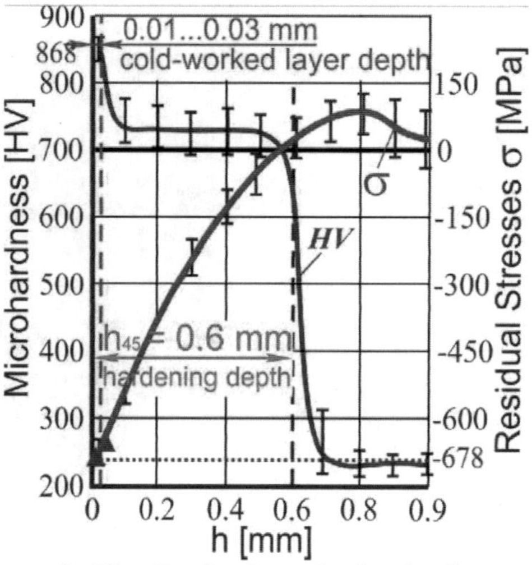

Figure 7. The distribution of microhardness and residual stresses in the workpiece surface layer after diamond smoothing.
▲ - residual stresses value acquired by X-ray technique.

4. Conclusions

It has been found experimentally that the suggested integration principle allows to perform workpiece processing on the same equipment thus ensuring minor values of stock allowance for the final stage of hybrid processing. This in turn ensures a significant decrease in the possible defects formation on the workpiece surface and contributes to the additional hardening of the workpiece. In the course of research it was registered that the integral processing leads to the surface microhardness increase by 12…17%, and the level of compressive residual stresses in the surface layer increase by 10…21% respectively.

Acknowledgements

This study was supported by a NSTU grant (project No. ТП-ПТМ-2_15)

References

[1] Zhu Z, Dhokia V G, Nassehi A and Newma S T 2013 *International Journal of Computer Integrated Manufacturing* **26** 596-615
[2] Makarov V M 2011 *RITM: Repair. Innovation. Technologies. Modernization* **64** 20-3
[3] Moriwaki T 2008 *CIRP Annals – Manufacturing Technology* **57** 736–49
[4] Skeeba V Yu, Pushnin V N, Erohin I A and Kornev D Yu 2015 *Materials and Manufacturing Processes* **30** 1408-11
[5] Nagae A 2004 *Proc. 11th Int. Conf. on Machine Tool Engineers (Tokyo)* pp 312-23
[6] Lauwers B, Klocke F, Klink A, Tekkaya A E, et al 2014 *CIRP Annals – Manufacturing Technology* **63** 561-83

IOP Conf. Series: Materials Science and Engineering **126** (2016) 012016 doi:10.1088/1757-899X/126/1/012016

[7] Skeeba V Yu 2015 *Actual Problems and Decisions in Machine Building* (Pfaffikon: Trans Tech Publ.)
[8] Ding H T, Shin Y C 2010 *International Journal of Machine Tools and Manufacture* **50** 106-14
[9] Brecher C, Hoffmann F, Karlberger A and Rosen C J 2008 *41st CIRP Conf. on Manufacturing Systems (Tokyo)* (London: Springer) pp 425-8
[10] Ivancivsky V V, Skeeba V Yu 2010 *Obrabotka metallov* **49** 15-21
[11] Skeeba V Yu, Ivancivsky V V, Zub N P and Turevich S V 2010 *The World of Scientific Discoveries* **2-3** 91-3
[12] Attia H, Tavakoli S, Vargas R and Thomson V 2010 *CIRP Annals – Manufacturing Technology* **59(1)** 83–8
[13] Salonitis K, Chondros T and Chryssolouris G 2008 *The International Journal of Advanced Manufacturing Technology* **38** 48-58
[14] Lobanov D V, Yanyushkin A S 2011 *Russian engineering research* **31** 236-9
[15] Pushnin V N, Erokhin I A, Kornev D Yu and Skeeba V Yu 2014 *Actual problems in machine building* **1** 245-55
[16] Pashkov E N, Martyushev N V and Masson I A 2014 *IOP Conference Series: Materials Science and Engineering*, **66 (1)** 012011
[17] Arkhipov P, Yanyushkin A, Lobanov D and Petrushin S 2013 *Applied Mechanics and Materials* **379** 124-30
[18] Pashkov E N, Martyushev N V and Yurovsky P G 2014 *Advanced Materials Research* **1040** 903-6
[19] Martyushev N V, Semenkov I V and Petrenko Y N 2014 *Advanced Materials Research* **872** 112-7.
[20] Plotnikova N, Losinskaya A, Skeeba V and Nikitenko E 2015 *Applied Mechanics and Materials* **698** 351-4
[21] Skeeba V Yu, Pushnin V N, Erohin I A and Kornev D Yu 2014 *Obrabotka metallov* **64** 90-102
[22] Brockhoff T 1999 *CIRP Annals – Manufacturing Technology* **48(1)** 255–60
[23] Vander Voort G F 2004 *ASM Handbook. Vol. 9: Metallography and microstructures* (Ohio: ASM International)
[24] Sharpe W N 2008 *Springer handbook of experimental solid mechanics* (Leipzig: Springer Science)
[25] Totten G E, Howes M and Inoue T 2002 *Handbook of residual stress and deformation of steel* (Ohio: ASM International)
[26] Ivancivsky V V, Skeeba V Yu and Pushnin V N 2011 *Obrabotka metallov* **53** 19-25
[27] Skeeba V Yu, Pushnin V N, Kornev D Yu and Parts K A 2015 *Obrabotka metallov* **68** 31-41
[28] Skeeba V, Ivancivsky V and Pushnin V 2015 *Applied Mechanics and Materials* **698** 288-93

Influence of Roller Burnishing Parameters on Depletion of Plasticity Reserve

V Yu Blumenstein [a], K P Petrenko [b]

28 Vesennyaya str., Kemerovo, 650000, Russia, T.F. Gorbachev Kuzbass State Technical University

[a] e-mail: blumenstein.vu@gmail.ru, [b] e-mail: konst.petr@mail.ru

Abstract. Roller burnishing process considerably increases surface quality and service life of machine parts. Efficiency of roller burnishing rises greatly when technological inheritance (TI) is taken into account. Research results of degree of plasticity reserve depletion (DPRD) while roller burnishing are presented. Results obtained made it possible to establish mechanisms of strain accumulation and plasticity reserve depletion according to roller burnishing parameters.

Introduction

Roller burnishing process is one of the most effective ways of work hardening. Scientific studies conducted until date indicate that parameters of roller burnishing properly assigned result in increasing fatigue resistance and service life of machine parts.

Processes of machining feature complex stress distribution and strain accumulation in surface layer. A large number of phenomena on tool-workpiece contact area that one should take into consideration restrain developing an all-purpose model of properties forming. At present, there are considerable quantity of studies that covers mechanics of machining [1-4].

F. Mohammadi, R. Sedaghati and A. Bonakdar developed a high fidelity 3D nonlinear finite element model (FEM) based on the isotropic hardening behavior to simulate low plasticity burnishing of titanium alloy [5]. In that model, the material properties not only depend on the strain, but also depend on the temperature and strain rate. As a result, the profitable compressive residual stress distribution in surface layer has been achieved.

I. Smolin, V. Kuznetsov, A. Dmitriev by using finite element method (FEM) resolved the problem relating stress-strain distribution while diamond burnishing. They concluded that under a certain conditions nanostructure formation in superficial layer is possible [6]. N. Chowdhury, A. Patwari, T. Rahman, S. Mostafa obtained analytical solution of elastic stress distribution in the vicinity of the rolling tool while roller burnishing [7].

F. Haji Aboutalebi and A. Banihashemi estimated the practical validation of Hooputra's ductile damage (HDD) criterion. The mentioned criterion is based on Kolmogorov's mathematical model and assumes that the equivalent plastic strain at the onset of damage is a function of stress triaxiality and strain rate [8]. The authors believe that HDD is a robust criterion for prediction of failure in ductile metals and that its parameters can be easily obtained from the forming limit diagram (FLD). Other damage criteria were also presented in the conditions of plastic deforming [9, 10].

A. Mahato, Y. Guo1, H. Yeung and S. Chandrasekar investigated metal surface flow in severe plastic deformation (SPD) by sliding using high-speed imaging [11]. The authors derived metal flow field and distribution of accumulated strain by the depth of surface layer. Yung-Chang Yen by using 2D and 3D modelling of ball burnishing process calculated strains in the surface layer and estimated surface quality parameters [12].

In up-to-date conditions, it is of great importance to numerically estimate accumulation of strain and forming of surface layer parameters from the attitude of technological inheritance (TI). A large number of works conducted until date indicate that above-mentioned approach can be implemented on the basis of TI mechanics [13]. The theory of quality TI at machining stages is based on the fundamental phenomenological model of continuous surface layer formation exposed to surface plastic deformation. Key mechanical parameters are degree of shear deformation Λ, indicator of stress state I, degree of plasticity reserve depletion (DPRD) Ψ.

Roller penetration and movement lead to the occurrence of a deformation site (DS) – local area of plastic flow in surface layer (fig. 1). When the surface layer is exposed to stressing, material particles move into the DS along flow lines *1*, *2* and *3*, plastic deformation reaches depth *h*. It results into the surface layer characterized with a various depth for shear deformation, plasticity reserve depletion, residual stress profile and surface properties formation. At the next machining operations, strain accumulation occurs with the regard of previously formed surface layer properties.

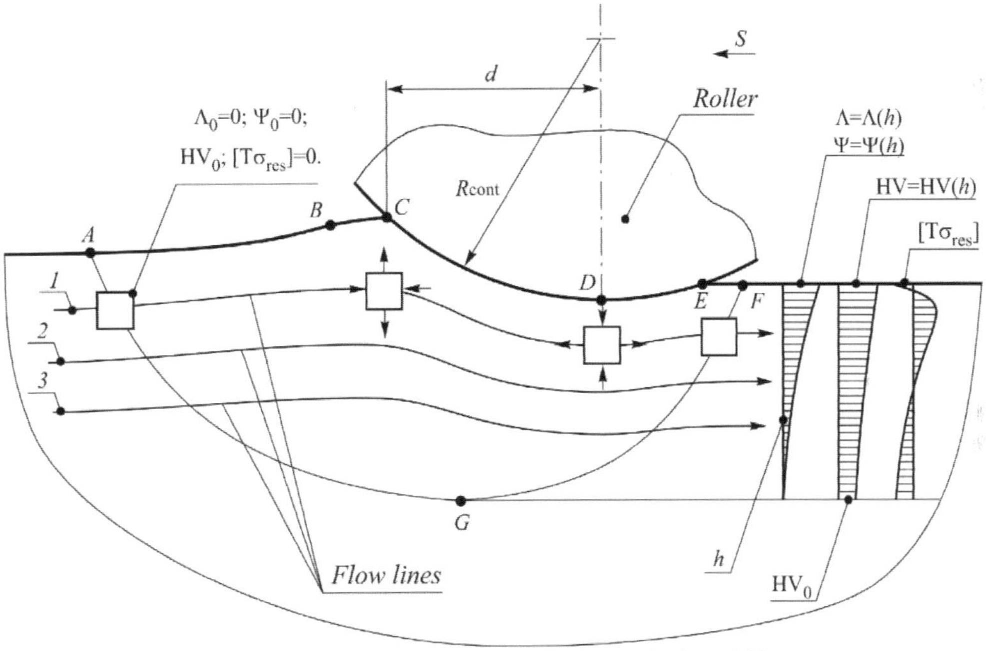

Fig. 1. Model of surface layer formation by roller burnishing process:
S – feed rate; R_{cont} – roller contour radius

Key feature of TI mechanics is stressing program (SP), which appears as variation of degree of shear deformation Λ with indicator of stress state I (fig. 2). LP comprises three steps of quasimonotonous deformation. On each step, continuous strain accumulation takes place in the conditions of changing indicator of stress state. At the key points of SP partial healing of defects and plasticity reserve recovery occur. SP which depends on roller burnishing parameters defines stress-strain state and forms surface layer properties. Under SP action in DS, plasticity reserve depletion takes place, which in turn results in surface layer properties forming.

The degree of plasticity reserve depletion was assessed as follows [14]:

$$\Psi = \Psi_1 + \Psi_2 = \int_0^{\Lambda_\kappa} \left[n\varphi_0 \Lambda_i^{n-1} + \left(1 - \varphi_o \Lambda_p^n (I)\right) \frac{1}{\Lambda_p (I)} \right] d\Lambda_i , \quad (1)$$

Ψ_1 – component dependent on flow stress or on accumulated deformation; Ψ_2 – component dependent on metal plasticity with I =const; Λ and Λ_p – accumulated and maximum permissible degree of shearing deformation with a specific stress state indicator I, n – strain-hardening coefficient; φ_0 – coefficient determined by plasticity tests. In the unstrengthened metal Ψ = 0, when plasticity reserve is completely depleted, Ψ = 1.

The advantage of above model is that it reproduces the nature on plastic deformation, which has two features: work hardening, characterized by accumulated strain and defects accumulation, specified by metal plasticity reduction.

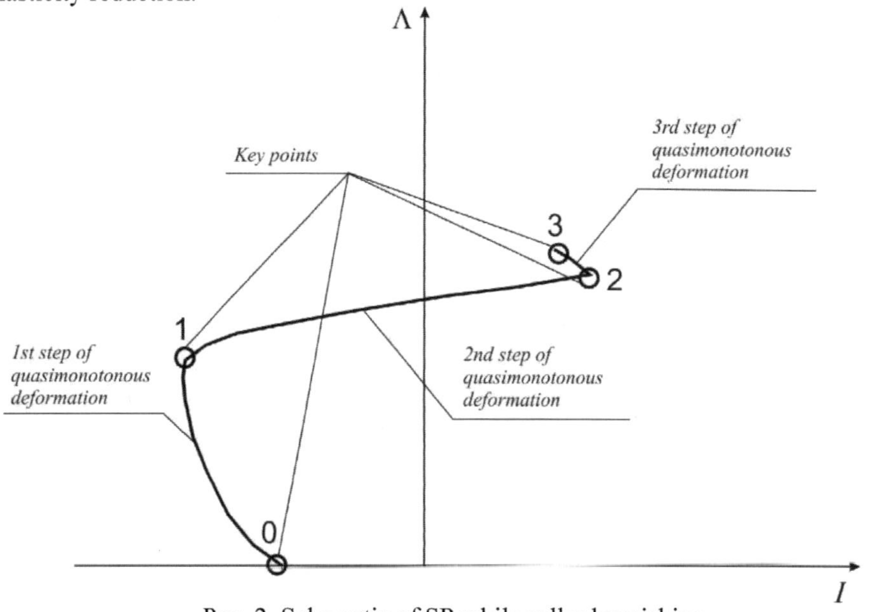

Рис. 2. Schematic of SP while roller burnishing

Materials and methods

In order to validate early obtained analytical models of SP and DPRD calculation experimental researches on roller burnishing of smooth round specimens made of 1045 carbon steel (table 1) were carried out. 1045 steel is one of the most widely used material in modern machine building industry, moreover, its plastic properties is well examined. Rollers were made of 52100 bearing steel.

Table 1: Chemical composition of 1045 steel, %

C	Si	Mn	Cr	S	P
			less than		
0.42-0.50	0.17-0.37	0.50-0.80	0.25	0.040	0.035

The burnishing parameters are presented in Table 2.

Table 2: Burnishing parameters

Specimen number	Number of passes until spalling	Roller diameter D_R, mm	Feed rate S, mm/rev	Spindle revolution rate n, min^{-1}	Roller contour radius R_{cont}, mm	Burnishing force P, N
51	3	95	0.07	630	2.5	2500
52	4				5.0	
54	3	64			8.5	2000
56	4				5.0	1500
57	5				13.5	2000

Specimens' surface layer was consistently burnished until complete plasticity reserve depletion and bulk material flaking off. Before every next pass, surface layer had fixed value of accumulated strain and DPRD. Burnishing on first pass was carried out almost at full specimen length with DS "freezing". Burnishing on second pass was performed on already machined surface and so on (fig 3).

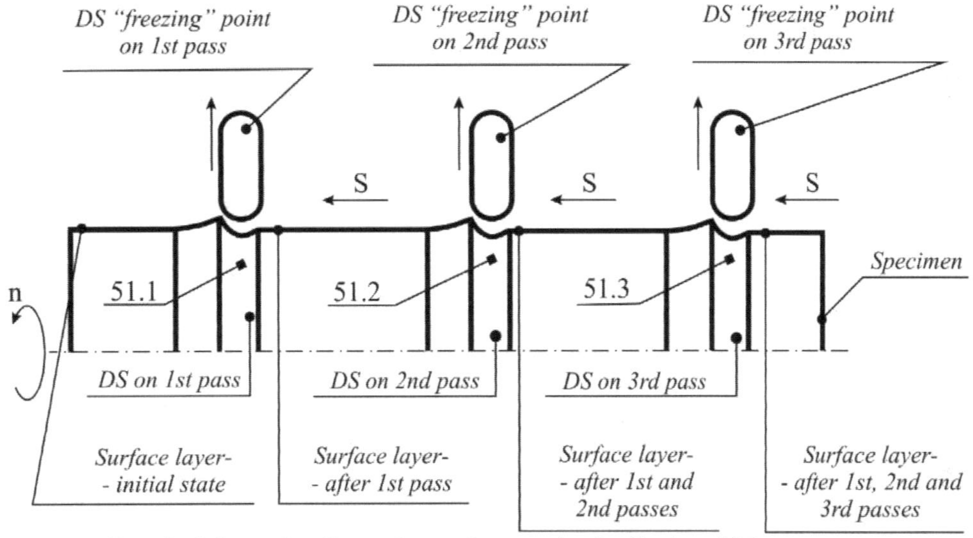

Рис. 3. Schematic of experimental research of roller burnishing process (specimen №51): S – feed rate, n – spindle revolution rate

Identification of sub-surface cracks, which correspond complete plasticity reserve depletion, was performed using scanning electron microscopy (SEM) on JEOL JSM-7500FA equipment (fig 4).

Geometric parameters of DSs obtained were recognized by using profile recorder "Talysurf 5M" by "Rank Taylor Hobson" company and were used as boundary conditions by FEM simulation. For that purpose, finite elements models of roller and work-piece were created (fig. 5).

Bottom edge of the work-piece was rigidly fixed. The indenter that was an absolutely rigid body was pressed and then shifted along horizontal line by the feed rate. There was no vertical displacement of the indentor. State of plane deformation was assumed. Initial conditions were Young's modulus E = 200 GPa, Poisson's ratio ν = 0.3, density ρ = 7800 kg/m^3, friction coefficient on the contact area f = 0.21 and flow curve that looked like as follows: $\sigma_i = 360 + 865.6\varepsilon_i^{0.57}$.

Рис. 4. Specimen surface layer at $\Psi = 1$, magnification × 50 000

Within the framework of above model nodal solution of stress-strain state parameters was obtained by using a special software system [15]. Then parameters of stress-strain state were recomputed along flow lines, general parameters of mechanics and SP were calculated. With knowledge of stress-state index in the selected point of flow line with the use of FLD maximum permissible the degree of shear deformation Λ_p was determined. Furthermore, instantaneous value of DPRD in the defined points of flow line was calculated. By adding up instantaneous values of DPRD, we received the accumulated value of DPRD along flow lines.

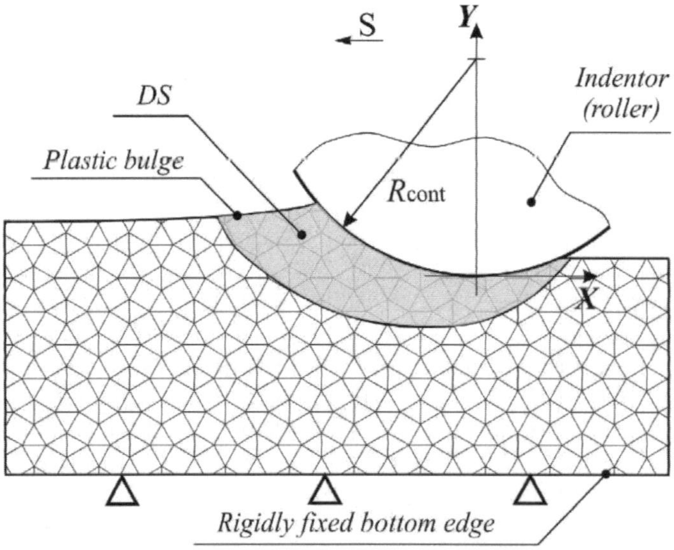

Рис. 5. FEM model for roller burnishing:
S – feed rate; R_{cont} – roller contour radius

Analytical calculation of DPRD was carried out at the defined point of flow lines by using previously obtained models of SP steps approximation which may be expressed as: $\Lambda = a_{ij}^{SP} \exp\left(b_{ij}^{SP} I\right) + c_{ij}^{SP}$ where $a_{ij}^{SP}, b_{ij}^{SP}, c_{ij}^{SP}$ – factors obtained from experimental researches [16].

Table 3 shows DPRD values, obtained by FEM and by analytical models.

Table 3: DPRD values obtained by FEM and by analytical models

Specimen and pass number	DPRD value, obtained by		DRPD error $\Delta\Psi$, %
	FEM Ψ_{FEM}	Analytical models Ψ_{AM}	
51.1	0.403	0.374	7.24
51.2	0.298	0.322	-8.55
51.3	0.256	0.275	-7.42
52.1	0.334	0.326	2.53
52.2	0.238	0.258	-3.01
52.3	0.223	0.224	-7.37
52.4	0.209	0.203	8.89
54.1	0.254	0.268	-5.51
54.2	0.241	0.233	3.31
54.3	0.211	0.211	-0.06
56.1	0.270	0.261	3.36
56.2	0.224	0.228	-1.94
56.3	0.209	0.205	2.00
56.4	0.176	0.190	-7.68
57.1	0.251	0.248	1.34
57.2	0.223	0.223	-0.13
57.3	0.190	0.205	-8.06
57.4	0.181	0.190	-5.04
57.5	0.172	0.180	-4.68

In Table 3 $\Delta\Psi$ – DRPD relative error defined as:

$$\Delta\Psi = \frac{\Psi_{FEM} - \Psi_{AM}}{\Psi_{FEM}} \cdot 100\% . \quad (2)$$

The variation of DRPD with burnishing force is shown in fig. 6. It is observed that DRPD increases on every pass with increase of the burnishing force. The variation of DRPD with roller contour radius is shown in fig. 7. It is clear that DRPD also increases on every pass with decrease of the roller contour radius. As shown in fig. 8, in every case DRPD increases with decrease of roller contour radius.

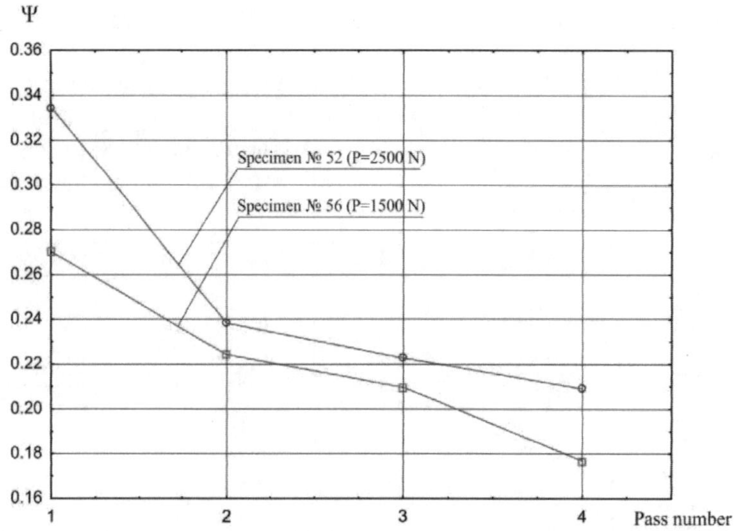

Fig. 6. Variation of DPRD with burnishing force (at a constant roller contour radius)

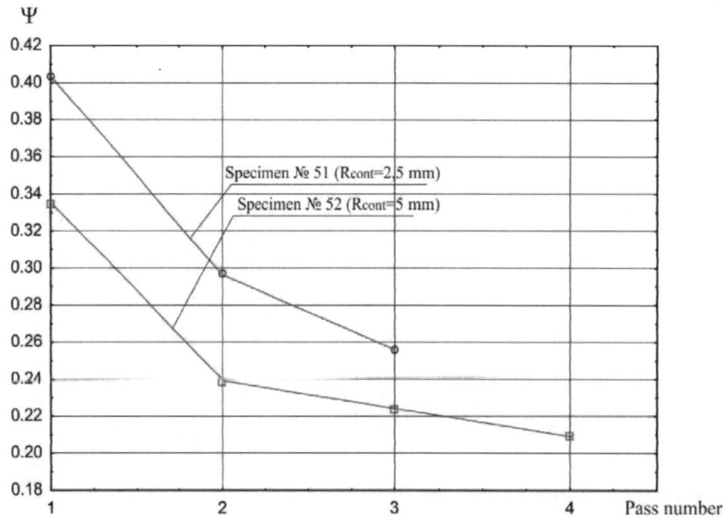

Fig. 7. Variation of DPRD with roller contour radius (at a constant burnishing force)

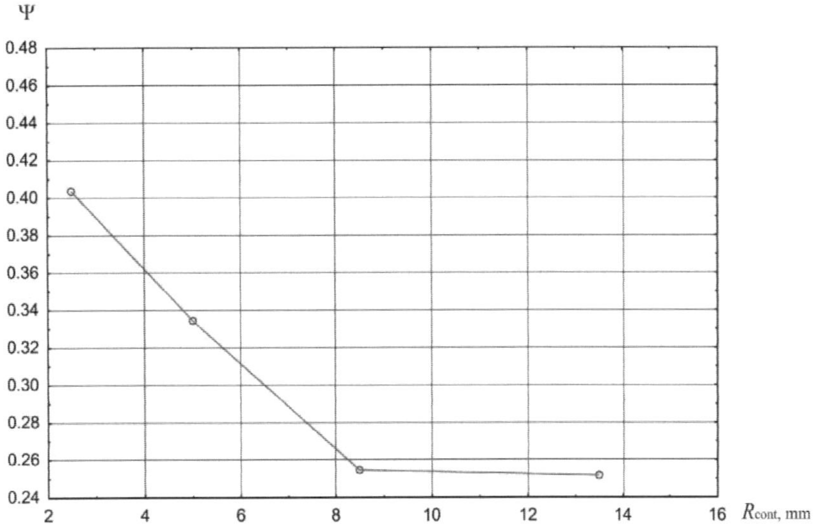

Fig. 8. Variation of DPRD on first pass with roller contour radius

Results presented in Table 3 show that DPRD values obtained by FEM correlate well with DPRD values obtained by analytical models. DPRP errors are less than 9 %.

Conclusions

1. Experimental research on roller burnishing of round specimens made of 1045 steel was conducted. Identification of sub-surface cracks was performed by using SEM. Geometric parameters of inherited DS were used as boundary conditions for FEM simulation.
2. Parameters of stress-strain state, SP and DPRD were calculated along flow lines in DSs.
3. It is found that decrease of roller contour radius and increase of burnishing force, which both lead to more heavy plastic flow in DS result in accumulated strain and DPRD growth.
4. Results obtained have shown the possibility of using geometrical parameters of inherited DS for estimation of degree of shear deformation, SP and DPRD of surface layer.

References

[1] J. Chakrabarty Theory of Plasticity, Third edition, Butterworth-Heinemann, 2006, 882 pp.
[2] P. L. B. Oxley, Mechanics of Machining, Ellis Horwood, Chichester (1989).
[3] E. G. Thomsen, C. Y. Yang, and S. Kobayashi, Mechanics of Plastic Deformation in Metal Processing, Macmillan, New York (1965).
[4] Smelyanskiy, V. M. Strengthening mechanics of machine part surface layer during surface plastic deformation, Moscow: MASHMIR, 1992, 60 p. (Rus)
[5] F. Mohammadi, R. Sedaghati, A. Bonakdar, Finite element analysis and design optimization of low plasticity burnishing process, Int. J Adv. Manuf. Techno. (2014), pp. 1337-1354.
[6] Igor Yu. Smolin Viktor P. Kuznetsov Andrey I. Dmitriev Numerical Analysis of the Stress-strain State of the Steel Superficial Layer under Nanostructuring Burnishing, XLII International Summer School Conference APM, 2014, pp. 468-474.
[7] N. A. Chowdhury, Anayet U. Patwari, T. Rahman, S. Mostafa Development of a Mathematical Model for Elastic Field Simulation During Unloading Condition in Surface Burnishing Processes Annals of the Faculty of Engineering Hunedoara –Journal of Engineering. Tome VI (year 2008), Fascicule 2, (ISSN 1584 – 266).
[8] F. Haji Aboutalebi, A. Banihashemi, Numerical estimation and practical validation of Hooputra's ductile damage parameters, Int. J Adv. Manuf. Technol. (2014), pp. 1701-1710.

[9] Y. Li, M. Luoa, J. Gerlachb, T. Wierzbicki, Prediction of shear-induced fracture in sheet metal forming, Journal of Materials Processing Technology. 210 (2010) pp. 1858–1869.
[10] J. Dzugan, M. Spaniel, P. Konopic, J. Ruzicka, J. Kuzelka, Identification of Ductile Damage Parameters for Austenic Steel, International Journal of Mechanical, Aerospace, Industrial, Mechatronic and Manufacturing Engineering Vol: 6, Num: 5, 2012, pp. 1034-1039.
[11] A Mahato, Y Guo1, H Yeung and S Chandrasekar Surface flow in severe plastic deformation of metals by sliding, 6th International Conference on Nanomaterials by Severe Plastic Deformation IOP Conf. Series: Materials Science and Engineering 63 (2014) 012016, pp. 1-8.
[12] Yung-Chang Yen Modeling of Metal Cutting and Ball Burnishing – Prediction of Tool Wear and Surface Properties – Dissertation –Presented in Partial Fulfillment of the Requirements for the Degree of Doctor of Philosophy in the Graduate School of the Ohio State University The Ohio State University 2004, p. 254.
[13] V.Yu. Blumenstein, Mechanics of technological inheritance during treatment and operation of machine parts / V.Yu. Blumenstein, V.M. Smelyanskiy, Mashinostroyeniye-1, Moscow, 2007, 400 pp.
[14] Yu.K. Filippov, Evaluation Criterion for the Quality of Parts Produced by Cold Forging//Forging and Stamping Production, 1999, № 2, P. 3-9.
[15] V.Yu. Blumenstein, Inherited Mechanic of Machine Parts Surface Layer / V.Yu. Blumenstein, A.A. Krechetov Official Registration of Computer Program № 2002619758. Application № 2002610463, Application Date 03.29.2002. Logged in Register of Computer Programs 05.18.2002.
[16] V.Yu. Blumenstein, Analytical Models of Stressing Programs at Stages of Machining / V.Yu. Blumenstein, K.P. Petrenko // Kuzbass State Technical University Bulletin, 2009, № 6, P. 77-83

IOP Conf. Series: Materials Science and Engineering **126** (2016) 012018 doi:10.1088/1757-899X/126/1/012018

Technological Aspects of Forming the Surface Microrelief of Low-Wear Coatings after Electro-Diamond Grinding

V G Burov, V V Yanpolskiy *, K Kh Rakhimyanov

Faculty of Mechanical Engineering and Technologies, Novosibirsk State Technical University, 20 Prospekt K. Marksa, Novosibirsk, 630073, Russia

*E-mail: yanpolskiy@corp.nstu.ru

Abstract. The results of electro-diamond grinding of coatings based on the WC25 powder material are presented in the paper. It is shown that after electro-diamond grinding of the WC25 coating, an obtained magnitude (Ra=2.02μm) of surface roughness doesn't meet the qualifying standards to parts surface working in wear-out conditions. The forming of the obtained microrelief is probably connected to the features of electrochemical dissolution of the WC25 coating material in the electrolyte being used. Based on the polarization studies, it is revealed that the electrochemical dissolution character of the indicated coating in the water solution of 10%$NaNO_3$ is determined by the dissolution character of cobalt (Co) component. The intensive cobalt (Co) dissolution during the electro-diamond grinding of the WC25 coating leads to the tungsten carbide chipping by the grinding disk particles that increases the roughness. One of the way to improve the surface quality of low-wear coatings after electro-diamond grinding is an introduction of an additional step in a technological process, carrying out with the switched off source of technological current. For realization of the process according to this scheme a technological dimension chain is made which takes into consideration the dissolution value of the most active coating composition element while the calculating of the operating dimensions of a detail.

1. Introduction

It is known that a failure of parts and construction elements is mainly caused by a wear and corrosive damage of their surface layers. The methods of a chemical-thermal impact and of surface hardening based on a deformation impact and also various ways of a low-wear coating application [1,2,3,4] are used to increase both the wear and corrosive resistance of parts surfaces. The application of such high-performance coating technologies as a detonation-gas and a plasma spraying, and an electro-beam surface welding allows to form the surface layers with high hardness [5,6,7,8]. The appearance of a new class of powder materials, consisting, as a rule, of a number of components and having a complex of unique properties, improves the technological processes of a preparation of backing material surface before coating and a final surface treatment after coating and welding operations. These technologies of the backing material surface preparation and the final machining influence greatly on the formation of the physical-mechanical characteristics and therefore, on the product reliability and longevity [9, 10]. As a rule, abrasive and diamond grinding are used for the final mechanical surface treatment after coating. It often results in the degradation of physical-mechanical surface layer characteristics obtained while the low-wear coating. The cause of the surface layer degradation is that hardness of the powder based coating is often comparable with abrasive tool hardness [11]. The significant perspectives for solving this problem are seen in using the technologies for low-wear coating treatment based on the combination of processes of electrochemical dissolution and mechanical cutting. The beneficial effects of using the specified technologies, providing the quality of the surface layer were obtained while forming the surfaces of the tough high-resistance steel alloys [12, 13].

2. Materials and procedure of the experiments

The experimental studies of the part machining with the powder based coating were carried out on an experimental facility for electro-diamond grinding (Figure 1).

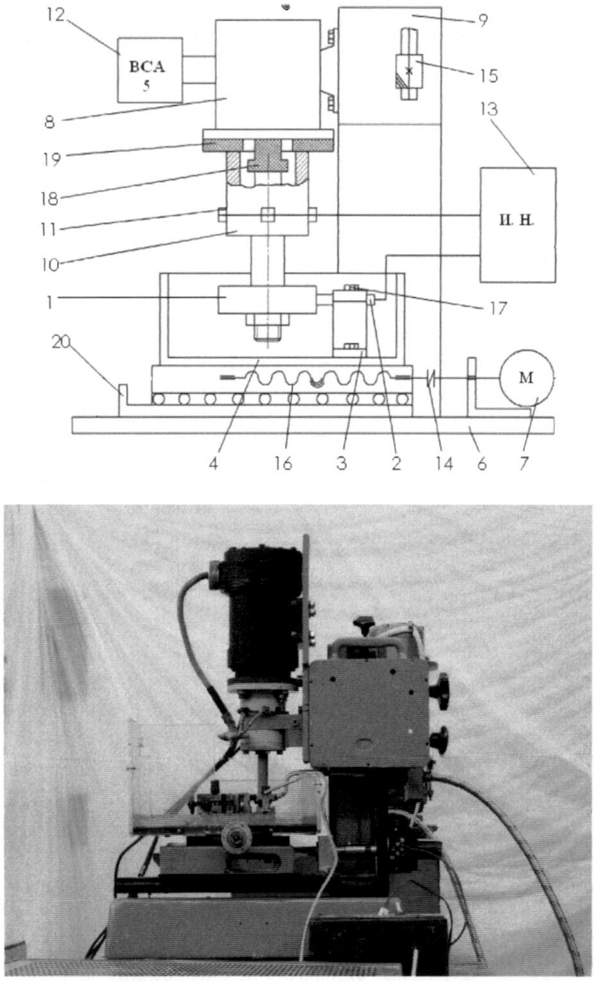

Figure 1. The scheme and appearance of an experimental facility for electro-diamond grinding. 1 – diamond disk on a metal alloy; 2 – test specimen; 3 – strut for fixing the test specimen; 4 – electrochemical cell; 5 – ball bearing ways; 6 – pedestal base;7 – feed motor; 8 – primary motor; 9 – carriage; 10 – spindle; 11 – contact-wiper facility; 12 – DC source–VSA 5; 13 – voltage source; 14 – joint box; 15 – cleat; 16 – lead screw; 17 – fixing screw; 18 – isolating joint box; 19 – textolite washer; 20 – cell limit stop.

The supplied technological voltage was U=8 V. The voltage limit is 8 V due to a probability of an erosion processes initiation while the electro-diamond grinding. It decreases the quality of the machined surface. A rate of motion was 15-17 m/c. The value of the longitudinal table feed was 45 mm/min. The diamond disk on a metal alloy ACB 80/63 100% M1 was used as a tool. The machining of the coated test specimens was carried out during one cut with the cutting depth t =0.1 mm.

A powder composition containing tungsten carbide is used as material for parts protection from abrasion wear [14]. The WC25 powder composition belongs to this material class. The coatings made of this composition have high wear resistance and they are used to restore movable and static joints. A detonation-gas sprayed coating based on the WC25 powder composition was chosen as the material for the research.

The quality of the machined surface was assessed by a roughness parameter. After the electro-diamond grinding the surface roughness of the coated test specimens was measured with the help of a complex New View 7300 Surface Profiler.

The research of electrochemical dissolution of the WC25 coating containing Cobalt (Co) and Tungsten (W) was carried out on potentiostat P5827M. The anode potential changed from 0 to 8 V. The water solution of neutral salt $NaNO_3$ was used as an electrolyte.

3. Results and discussion

In the results of the experiments on the electro-diamond grinding of the coatings based on the WC25 powder composition it was found that the obtained surface roughness magnitude Ra=2.02μm (Figure 2). It should be noted that higher quality demands are made to the working surfaces, specifically to the surface roughness, of the coated parts relating to a wear-resistant class.

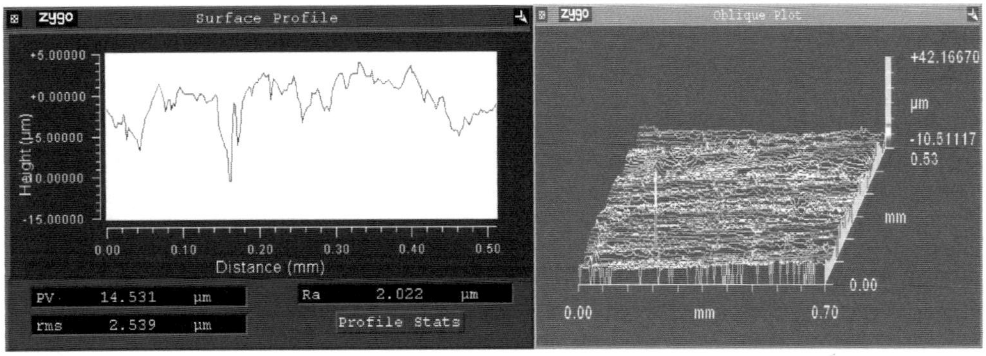

Figure 2. The surface roughness of the test specimen with the WC25 coating after electro-diamond grinding

Probably, the forming of the microrelief received after electro-diamond grinding is caused by the features of the electrochemical dissolution of the WC25 coating material in the electrolyte used. These features are stipulated by the different behavior of the WC25 components, tungsten carbide and cobalt alloy, during the electrochemical dissolution. Specifically, an active cobalt (Co) dissolution in the water solution of 10% $NaNO_3$ occurs in the potential range from $\varphi = 0$ V to $\varphi = 8$ V, which is indicated by a current density increase with the anode potential rise. (Figure 3)

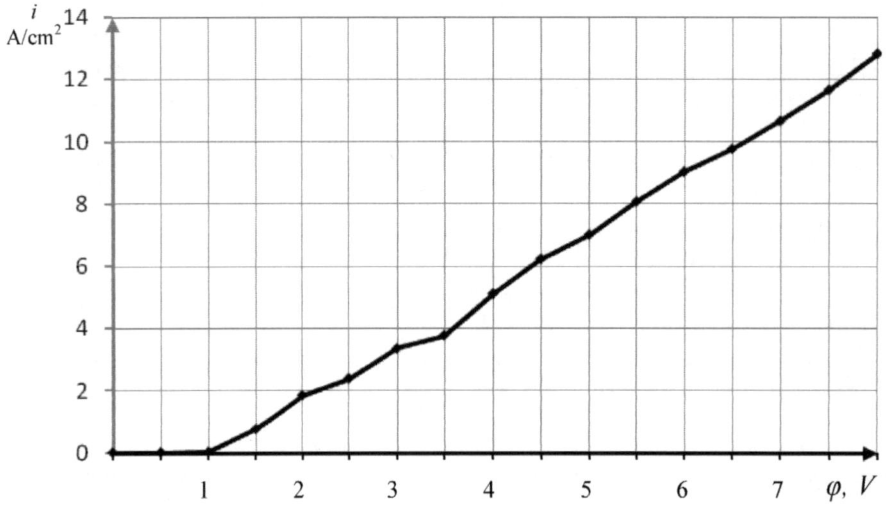

Figure 3. Polarization curves of anode dissolution of cobalt (Co) in water solution of neutral 10%$NaNO_3$

The feature of tungsten (W) electrochemical behavior in the water solution of 10% $NaNO_3$ is that the dissolution in an active state is only observed until the potential φ = 1.5 V (Figure 4). Further anode potential rise results in the current density decreasing practically till 0. Probably, in the potential range from 1.5 V to 8 V it is caused by the formation of an oxide film (Figure 5) on the anode surface in the result of tarnishing. This film retards the anode dissolution and results in a surface passivation.

The analysis of the received results of electrochemical dissolution of cobalt (Co) and tungsten (W) allows to make an assumption that the character of the anode behavior of the WC25 coating in the water solution of 10% $NaNO_3$ (Figure 6) is determined by the character of cobalt (Co) component dissolution.

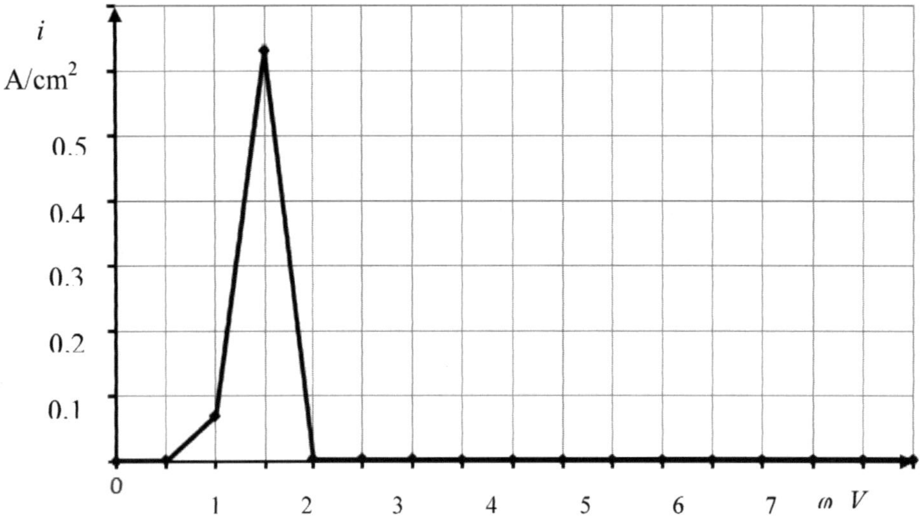

Figure 4. Polarization curves of anode dissolution of tungsten (W) in water solution of 10% $NaNO_3$

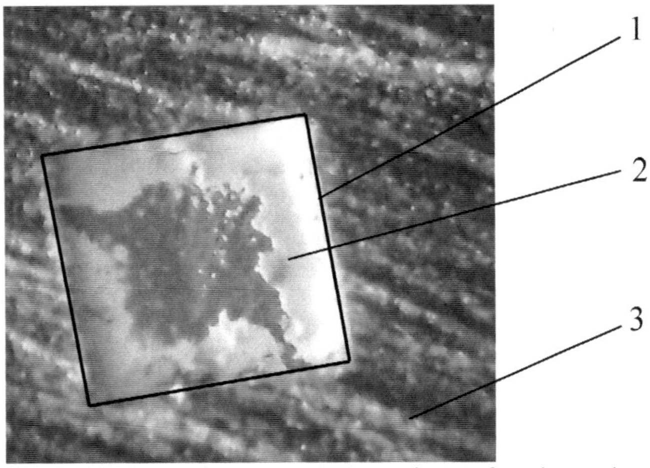

Figure 5. Surface morphology of tungsten (W) specimen after electrochemical dissolution in 10% $NaNO_3$ water solution: 1 – specimen; 2 – passivating film; 3 - arbor

Thus, we can observe an intensive dissolution of cobalt (Co) in the water solution of 10% $NaNO_3$ which causes the exposure of the particles of tungsten carbide and results in their chipping out during the process of cutting by a diamond grinding disk. The similar mechanism of forming the surface microrelief was observed while the electro-diamond grinding of hard alloys [15].

Probably, the value reduction of surface roughness after electro-diamond grinding is possible by the choice of a passivating electrolyte composition. However, the application of a similar variant for the reduction of surface roughness in case of the electro-diamond grinding of low-wear coatings based on powder compositions is problematic because of the complexity of a composition choice of a passivating electrolyte for all elements.

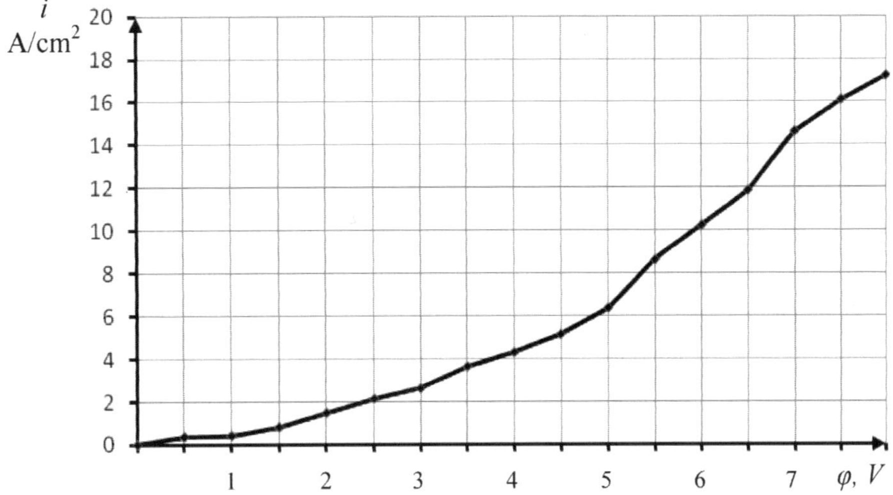

Figure 6. Polarization curves of anode dissolution of the VK25 powder composition in a water solution of 10% $NaNO_3$

One of the way to improve the surface quality of low-wear coatings after electro-diamond grinding is an introduction of an additional step in the technological process, carrying out with the switched off source of technological current. However, for realization of the process according to this scheme it is necessary to take into consideration the dissolution value of the most active composition element while calculating the operating dimensions of a detail. A technological dimension chain made up, subject to a possible dissolution value of the coating surface based on the WC25 after electro-diamond grinding is shown in Figure 7.

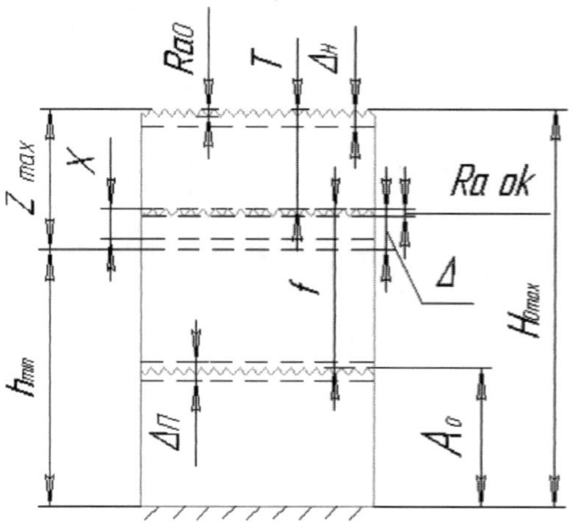

Figure 7. Technological dimension chain for a value definition of the minimal total grinding amount for machining the WC25 coating, subject to a possible value of dissolution of the coating surface after electro-diamond grinding.
H_{0max} – maximum coating dimension before machining; Z_{max} – maximum machining allowance; Δ – allowance for a final part dimension; h_{min} – minimal coated part thickness after machining; ΔH – allowance for coating thickness before machining; ΔП – thickness of a transition layer ; A_0 – thickness of a backing material; f – minimal sufficient coating value for part exploitation; Ra_0 – coating surface roughness before machining; Ra_{0K} – final surface roughness X – value of the coating surface dissolution after electro-diamond grinding.

3. Conclusion

These studies showed that an increased surface roughness value (Ra=2.02μm) is observed after the electro-diamond grinding of the coating based on the WC25 powder composition. The forming of the surface microrelief after electro-diamond grinding is connected to the features of the electrochemical dissolution of the WC25 coating elements in the electrolyte under study. The character of the dissolution of the indicated coating is determined by the character of the electrochemical dissolution of cobalt (Co). We observe an intensive dissolution of cobalt (Co) in the water solution of 10% $NaNO_3$ which causes the exposure of the particles of tungsten carbide and results in their chipping out during the process of cutting by a diamond grinding disk. To improve the surface quality of low-wear coatings after electro-diamond grinding is possible by an introduction of an additional step in the technological process, carrying out with the switched off source of technological current.

References

[1] Iskhakova, G.A., Rakhimyanov, Kh.M., 1987 Investigation of the microstructure and mechanical properties of steel 45 after plasma hardening, *Soviet surface engineering and applied electrochemistry* **5** 28-32

[2] Vadim, S. Valery, P., Ivan, E., Dmitry, K., 2015 Integration of production steps on a single equipment, *Materials and Manufacturing Processes* Vol. **30** 1408-11

[3] Rakhimyanov, Kh.M., Iskhakova, G.A., Karmanov, L.L., Grodnikas, G.Kh., 1993 The influence of ultrasonic deformation on the steels surface lay formation by sparking alloying with cemented carbide electrode, *Elektronnaya Obrabotka Materialov*. **1** 17-20

[4] Dudina, D.V., Batraev, I.S., Ulianitsky, V.Y., Bulina, N.V., Korchagin, M.A., Bataev, I.A., Jorge, A.M., 2014 Formation Routes of Nanocomposite Coatings in Detonation Spraying of Ti3SiC2-Cu Powders, *Journal of Thermal Spray Technology* **23** 1116-23

[5] Dudina, D.V., Korchagin, M.A., Zlobin, S.B., Ulianitsky, V.Y., Lomovsky, O.I., Bulina, N.V., Bataev, I.A., Bataev, V.A., 2012 Compositional variations in the coatings formed by detonation spraying of Ti 3Al at different O 2/C 2H 2 ratios, *Intermetallics* **29** 140-6

[6] Dudina, D.V., Mali, V.I., Anisimov, A.G., Bulina, N.V., Korchagin, M.A., Lomovsky, O.I., Bataev, I.A., Bataev, V.A., 2013 Ti3SiC2-Cu composites by mechanical milling and spark plasma sintering: Possible microstructure formation scenarios, *Metals and Materials International* **19** 1235-41

[7] Galchenko N.K., Kolesnikova K.A., Belyuk S.I., Dampilon B.V., 2014 Structure and properties of boride coatings synthesized from thermo-reactive powders during electron-beam surfacing, *Advanced Materials Research* **880** 265-71

[8] Ruktuev A.A., Golkovski M.G., Samoylenko V.V., Komarov P.N., Bataev I.A., Bataev A.A., 2014 Corrosion resistance of multilayer Ti-Ta coatings obtained by electron beam cladding in the atmosphere, *Advanced Materials Research* Vol. **1040** 759-63

[9] Rakhimyanov, Kh.M., Iskhakova, G.A., Rakhimyanov, A.Kh. 1996, The role of ultrasonic plastic deformation in spark alloying, *Fizika i Khimiya Obrabotki Materialov* **1** 68-72

[10] Iskhakova, G.A., Gileta, V.P., Rakhimyanov, Kh.M. 1991, Structure and mechanical properties of WC-Co alloy surface layer after diamond-ultrasonical treatment, *Sverkhtverdye Materialy* **5** 54-61

[11] Kremen', Z.I., Lebedev, A.I. 2011, High-porosity wheels based on cubic boron nitride for defect-free grinding, *Russian Engineering Research* **31** 867-9

[12] Mogilnikov, V.A., Chmir, M.Ya., Timofeev, Yu.S., Poluyanov, V.S. 2013, Diamond - ECM grinding of ceramic-metal tungsten, *Procedia CIRP* Vol. **6** 407-9

[13] Arkhipov P.V., Yanyushkin A.S., Lobanov D.V., Petrushin S.I. 2013The effect of diamond tool performance capability on the quality of processed surface, *Applied Mechanics and Materials* Vol. **379** 124-30

[14] Borisov, Yu.S., Kolisnichenko, O.V. 2003, Influence of the conditions of item surface heating on the structure of strengthened layers of U8 steel in plasma-detonation treatment, *Avtomaticheskaya Svarka* **3** 31-5

[15] Lobanov, D.V., Yanyushkin, A.S. 2011, Influence of sharpening on the quality of hard-alloy tools for the cutting of composites, *Russian Engineering Research* **31** **(3)** 236-9

Residual Stress, Structure and Other Properties Formation by Combined Thermo-Hardening Processing of Surface Layer of Gray Cast Iron Parts

Kh M Rakhimyanov, Yu V Nikitin, Yu S Semenova and A S Eremina

Faculty of Mechanical Engineering and Technologies, Novosibirsk State Technical University, 20 Prospekt K. Marksa, Novosibirsk, 630073, Russia

E-mail: kharis51@mail.ru

Abstract. The proposed combined thermo-hardening processing of gray cast iron enables to control the surface layer structure and mechanical properties formation. The processing includes high-speed heating by low-temperature plasma source and ultrasonic surface plastic deformation. The algorithm of calculation the stress-strain state of a surface layer at combined processing of gray cast iron is developed. This algorithm is based on method of sections. The ultrasonic surface deformation contribution is determined during formation of residual stresses. It is established that the combination of the thermal and deformation effects on the material provides an additional increment of microhardness and increase of surface layer thickness. Experimental results shows that the features of structural and phase transformations in a surface layer are revealed without a surface melting by energy of low-temperature plasma. The top of a layer does not contain inclusions of graphite that testifies to change of structural transformations in conditions of combined processing.

1. Introduction

The durability of machine parts working in conditions of alternating load, fatigue wear, high dynamic load is determined by mechanical properties and structure of the material. The gray cast iron has the rather low strength characteristics, so the formation of favorable stress-strain state condition of a surface layer of machine parts is important.

Improving performance properties of the surface layer can be achieved using thermal and deformation hardening methods. Efficiency of laser and plasma hardening methods are shown by authors [1, 2]. Results of hardening by deformation, including ultrasonic methods are presented in papers [3, 4, 5]. However, combined technologies based on a combination of high thermal and deformation effects represent significant opportunities to provide high-strength state of the surface layer [6].

The surface layer properties during combined thermo-hardening processing are formed in two stages: ultrasonic surface plastic deformation and high-speed heating by energy of low-temperature plasma. Each of this stage changes the structural and stress-strain state of the material.

The ultrasonic surface plastic deformation is carried out by the instrument oscillating with an amplitude 2A at static load P (Figure 1). Thermal and structural changes caused by high-speed heating by a source of density of power q with a concentration factor k (Figure 1). For effective control of processing the study of structural and stress-strain state forming is necessary.

However mechanical and X-Ray methods of residual stresses determination allows studying only final condition. Mechanical methods as a rule have destructive testing character. They are not applicable for working machine parts and requires manufacturing of special pattern. Furthermore the X-ray method allows to determine stress only in the limited surface layer. Thus mathematical modeling which aimed at simulation of stresses and deformations forming process at combined effect is of interest.

Structural and phase transformations were investigated by metallographic analysis. The microhardness was accepted as a characteristic of mechanical properties of a surface layer.

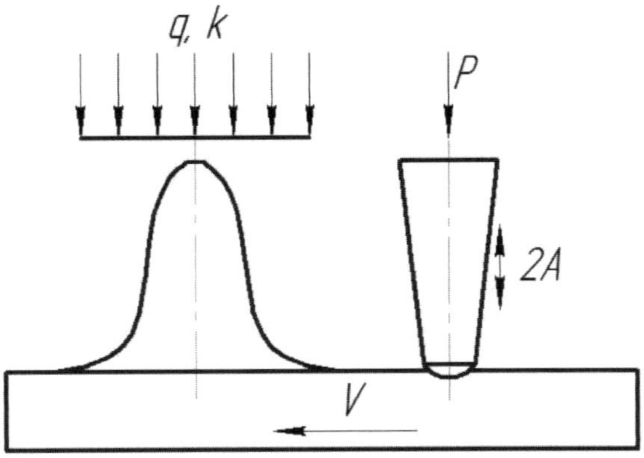

Figure 1. Scheme of the combined processing.

2. Materials and methods

A well-known method of sections was applied to calculate the stress-strain state [6]. In this paper the earlier moment of calculation for combined processing when the ultrasonic deformation precedes high-speed heating is not shown. In this research processing presented as combination of two stages - deformation and thermal effects. At a both stages the scheme of interaction of material with a source of effect can be approximated by the scheme of an equivalent rod system. According to a method of sections we can split a part on computational layers. There is condition that within the limits of each layer it is possible to consider deformations, stresses and temperatures constant for current step of calculations.

The rod approximation allows to determine stresses in computational layers from one layer to another though the method of series tests and iterations under the formula:

$$\sigma_{zi} = \left[-\delta_i + e_z + \Theta_x y_i\right] E_i \quad (1)$$

where δ_i - a total strain of a calculated layer; e_z - relative axial strain; Θ_x - relative angle of rotation of transversal sections; y_i - the ordinate of a computational layer center; E_i - coefficient of elasticity.

At the first stage of processing the surface layer is formed by imposing impressions from separate imprints of ultrasonic tool. The intensity of deformation in each imprint within the limits of depth h_s is determined by expression:

$$\varepsilon_i = \varepsilon_{i,0} \exp\left[-C \cdot \frac{y_i}{h_s}\right] \quad (2)$$

Depth of distribution of plastic deformation (h_s) and it's intensity in center of an imprint ($\varepsilon_{i,0}$) can be expressed through parameters of elastic-plastic contact of the tool with a surface:

$$\varepsilon_{i,0} = 2.4 \cdot \sqrt[3]{\left(\frac{2hk_d^{-2}}{D_p}\right)^2 \cdot \left(\frac{B}{A}\right)^{0.2}}; \quad (3)$$

$$h_S = k_d \sqrt{9.81 D_p h_{\max} - 1.42ab}, \quad (4)$$

where k_d, D_p, A, B, a, b - the geometric parameters of a contact. This parameters are determined by the solution of differential equation of the tool motion with allowance for dynamic properties of gray cast iron [6]; h_{max} - maximum depth of the ultrasonic tool indentation and h - depth of residual the plastic imprints.

Out the plastically deformed zone we accept the linear law of distribution of intensity of deformation:

$$\varepsilon_i = k_\varepsilon y_i \tag{5}$$

where k_ε - constant of proportionality, defined of the equilibrium condition of internal forces.

Applying rod approximation we consider, that the intensity of deformation is numerically equal to the biggest deformation. Receiving in (1) $|\delta_i|=|\varepsilon_i|$ we can define stresses in computational layers.

Summarizing values of deformations and stresses from a step to step during process of deformation we can obtain the characteristics of the stress-strain state to the end of the first stage of processing:

$$\delta_{R,i} = \sum_k \delta_{i,k}; \qquad \sigma_{R,i} = \sum_k \sigma_{i,k}; \qquad e_{Z,i} = \sum_k e_{Z,k}; \quad \Theta_{x,R} = \sum_k e_{x,k} \tag{6}$$

where k - number of a step of calculation.

At the second stage of processing the computational layers will be determined by the sum of thermal expansions of graphite inclusions (δ_G) with temperature and structural extensions of a metal matrix of gray cast iron ($\delta_{MT}+\delta_{MS}$):

$$\delta_i = \alpha\delta_G + \beta(\delta_{MT} + \delta_{MS}) \tag{7}$$

where α and β - volumetric share of graphite inclusions and metal matrix (for gray cast iron $\alpha=0.025$, $\beta=0.975$).

The matrix strains were defined through changes of volumes of phases, which are included in it's structure:

$$(\delta_{MT} + \delta_{MS}) = \frac{V_T - V_{T_0}}{3V_{T_0}} \tag{8}$$

where V_T and V_{T_0} - volumes of a matrix at initial (T_0) and current (T) temperature.

In it's turn volume of a matrix at current temperature can be calculated as the sum of volumes of separate phases (V_{jT}):

$$V_T = \sum_j \alpha_j V_{\alpha_j T} \tag{9}$$

where a_j - volumetric contents of jth phase in metal matrix.

According to the diagram of volumetric condition of phases the temperature relations of volumes of possible phases included in the matrix of the gray cast iron (ferrite (V_α), cementite (V_c), austenite (V_γ), martensite (V_M)) can be calculated by expressions:

$$\left.\begin{array}{ll} V_\alpha(T) = 0.12708 + 5.528\cdot10^{-6}T; & V_\gamma(T,C) = 0.12282 + 8.56\cdot10^{-6}T + 2.15\cdot10^{-3}C; \\ V_c(T) = 0.13023 + 4.884\cdot10^{-6}T; & V_M(T,C) = 0.12708 + 4.448\cdot10^{-6}T + 2.79\cdot10^{-3}C \end{array}\right\}, \tag{10}$$

where C - percentage of carbon in a phase.

Volume of matrix (V_T) is determined pursuant to regularities of structural transformations at heating and cooling for gray cast iron (Table 1). In Table 1 the following labels are adopted:

A_{c1} - temperature of the beginning of austenite transformation determined with allowance for influence of silicon, included in cast iron a structure, and high-speed character of heating;

V_p - volume of a perlite matrix of gray cast iron:

$$V_T = V_P = 0.88\cdot V_\alpha(T) + 0.12\cdot V_c(T) \qquad T\in[T_0; A_{C1}] \tag{11}$$

where 0.88 and 0.12 - volumetric shares of ferrite and cementite respectively;

A_γ - volumetric share of austenite, generated to a moment, when temperature increase up to current value T. The research on a kinetics of α-γ transformation at continuous high speed heating and data obtained from the solution of a diffusion problem, show that duration of transformation medium laminated perlite in austenite has exponential character. It can be calculated by expression:

$$\tau_{\alpha-\gamma} = 0.1 \cdot \exp(-0.00051 \cdot (V_H - 10^3)) \qquad V_H \in \left[10^3;10^4\right] \tag{12}$$

At known values of heating speed (V_H) and time $\tau_{\alpha\text{-}\gamma}$ the temperature T_k at which the transformation will be finished, can be determined by expression:

$$T_k = A_{C1} + V_H \cdot \tau_{\alpha-\gamma} \tag{13}$$

Accepting the exponential low of change A_γ,and taking to account that $T=T_k$ and $A_\gamma=1$, we can present change A_γ during heating by expression:

$$A_\gamma = \left(1 - \exp\left[-6.9\frac{T - A_{C1}}{T_K - A_{C1}}\right]\right) \qquad T \in \left[A_{C1};911°C\right] \tag{14}$$

$$M = \left(\frac{0.05(M_H - T)}{M_H - 20}\right)^{0.2} \qquad M_K \le T \le M_H \tag{15}$$

where M_H andM_k - temperature of a beginning and the end of martensite transformation.
M- the volumetric share of martensite, generated to a moment, when temperature downturn up to current value T. Taking into account, that austenite is transformed in martensite, near to temperature M_H and that at surface quenching of gray cast iron in a structure of a hardened layer remains from 40 up to 60% residual austenite, we can obtain M by expression:

Table 1. Values of specific volume of matrix (V_T) of gray cast iron

	$T_{max} < A_{c1}$	$A_{c1} \le T_{max} < 911°C$	$T_{max} \ge 911°C$
Heating	$T < A_{c1}$	$A_{c1} \le T < 911°C$	$T \ge 911°C$
	V_P	$(1 - A_\gamma)V_P + A_\gamma V_\gamma$	V_γ
Cooling		$M_H < T \le T_{max}$	
	V_P	$(1 - A_{\gamma max})V_P + A_{\gamma max}V_y$	V_γ
		$M_K \le T \le M_H$	
		$(1 - A_{\gamma max})V_P + A_{\gamma max}\left[(1 - M)V_\gamma + MV_M\right]$	$(1 - M)V_\gamma + MV_M$

To unit calculations of stresses and deformations at two stages in process of combined processing simulation the fulfillment of the following requirements is necessary:

1. At both stages the calculations should be executed for the same computational layers;
2. Resultant values of the characteristics of the stress-strain state obtained on expression (6) after ultrasonic deformation should be introduced as initial values in the beginning of second stage of calculation.

At the experimental part of research the samples with sizes 25x25x100 mm of gray cast iron, containing C-3.3%, Si-2.3%, Mn-0.46%, Cr-0.092%, Ni-0.12% are used. Ultrasonic deformation execute by the tool with spherical working part (indenter) with radius 4mm. Oscillation amplitude accepted 15 μm and oscillation frequency 20 kHz. For plasma heating the plasma generator with a tungsten electrode and fixation of arc length was used.

The modes of ultrasonic deformation and plasma heating were set from condition of obtaining of hardened layer with maximum thickness and absence of surface melting.

3. Results and discussion

The simulation of the first stage of processing has shown, that the ultrasonic deformation on all investigated modes (amplitude from A=5 up to A=15μm, static effort from P=50 up to P=1500 N, the ultrasonic frequency from f=18 up to f=66 kHz) forms in a surface layer compressing residual stresses. Distribution of residual stresses on a thickness of detail is shown on Figure 2.

The experimental data shows that the ultrasonic plastic deformation creates a surface layer by a thickness up to 300 μm with microhardness up to 3560 MPa for gray cast iron with initial structure containing up to 95% of perlite and inclusions of flaked graphite and with microhardness 1850-2100 MPa (Figure 3, curve 1).

The analysis of the deformed layer structure has shown that such increase of microhardness is stipulated first of all by deformation of a matrix. The ultrasonic processing results to increas of dispersibility of matrix structural components - splitting, deforming of cementite plates in perlite (Figure 4). The surface roughness after ultrasonic deformation redused from Ra 2.5 μm to Ra 0.66 μm. The completely new regular microrelief was generated.

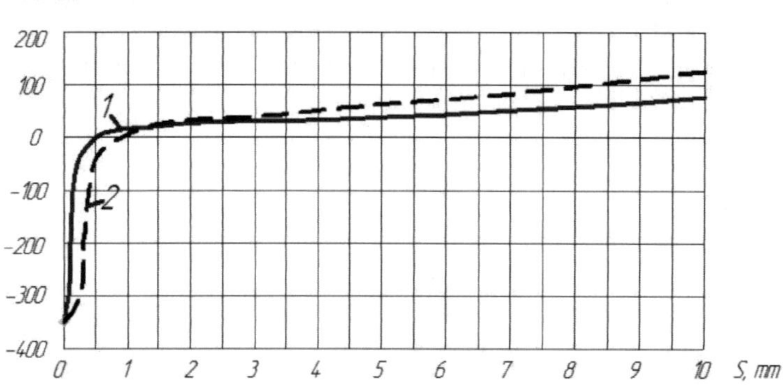

Figure 2. Distribution of residual stresses on a thickness of detail (f=18 kHz, D=8 mm, 1 - P=185 N, A=5 μm, 2 - P=420 N, A=15 μm)

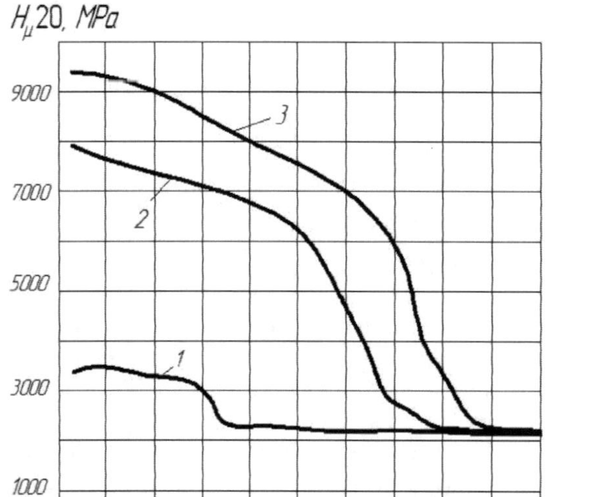

Figure 3. Distribution of microhardness on depth of a surface layer:
1 - ultrasonic deformation; 2- plasma hardening; 3 - combined processing

IOP Conf. Series: Materials Science and Engineering **126** (2016) 012019 doi:10.1088/1757-899X/126/1/012019

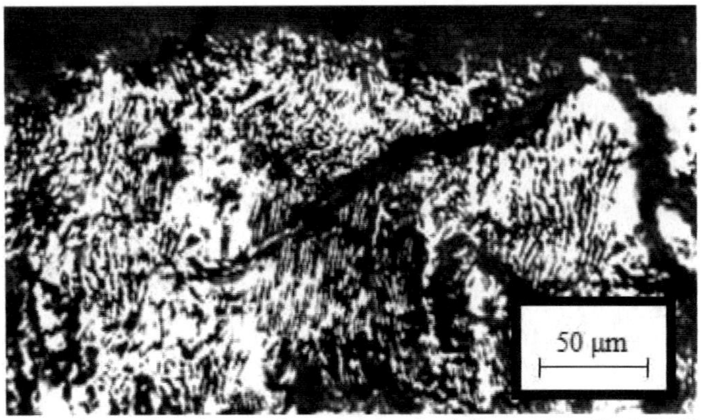

Figure 4. Structure of a surface layer after ultrasonic deformation (P=250 N, V=50 m/min)

At simulation of the second stage of processing it is established that in computational layers testing phase transformations at a stage of heating stresses reach a yield point. Therefore stresses changes only in layers, in which phase transformations does not happen. Change of modes of thermal effect (q, k, V) reducing depth of the phase transformation penetration and increase influence of ultrasonic deformation on a state of material at combined processing. Thus, changes of a combination of ultrasonic deformation modes with modes of high-speed heating allow to control process of stresses and deformations forming.

The high-speed surface heating by energy of low-temperature plasma without surface melting allowed to get a surface layer with thickness of 600 μm and microhardness up to 7500 MPa (Figure 3, curve 2).

For determination of graphite inclusions behavior during thermal processing the following experiment was conducted. Two samples enclosed in metal holders with microsections were prepared and microphotos of initial structure made at increase x500. Then in a special device the surfaces of microsections were densely pressed to each other. In such position the outside surfaces of samples were thermally treated. Then the microphotos of face surfaces were again maid at the same increase. The comparison of the shape and sizes of graphite inclusions before and after heat treatment has not revealed essential changes. Apparently it is possible to explain this fact by a narrow time-temperature interval of austenite transformation in gray cast iron, because it is insufficient for dissolution of large graphite inclusions.

The analysis of a structure after etching by a 4% alcoholic solution of nitric acid has shown that it is possible to select different zones in it (Figure 5). At all zones there are graphite inclusions. The metal matrix in a zone 1 (40 μm in depth) is formed by structures of a ledeburite. In a zone 2 (about 80 μm) the matrix has a structure of a short-acicular martensite. The zone 3 represents a transient layer (60-80 μm) consisting from short-acicular and unstructured martensite. The zone 4 has the greatest expansion (300 μm) and the fine structure, which can be classified on a level of hardness as an unstructured martensite.

The zone 5 (100 μm) is formed by a sorbitic layer which hardness is gradually reduced and on a distance 600 μm from a surface reached values of initial structure characteristics.

The combined processing allowed to increase hardness (up to 9300 MPa) and thickness of surface layer (up to 750 μm) (Figure 3, curve 3). The analysis of microstructure has shown that the zone 1 with a ledeburite structure has the larger sizes than at simple plasma heating and does not contain graphite inclusions (Figure 6). As the modes of thermal effect did not vary, it can be assumed that such difference in a structure of zones caused by change of a thing structure of graphite and metal matrix during ultrasonic deformation, that probably is reflected on kinetics of phase transformations.

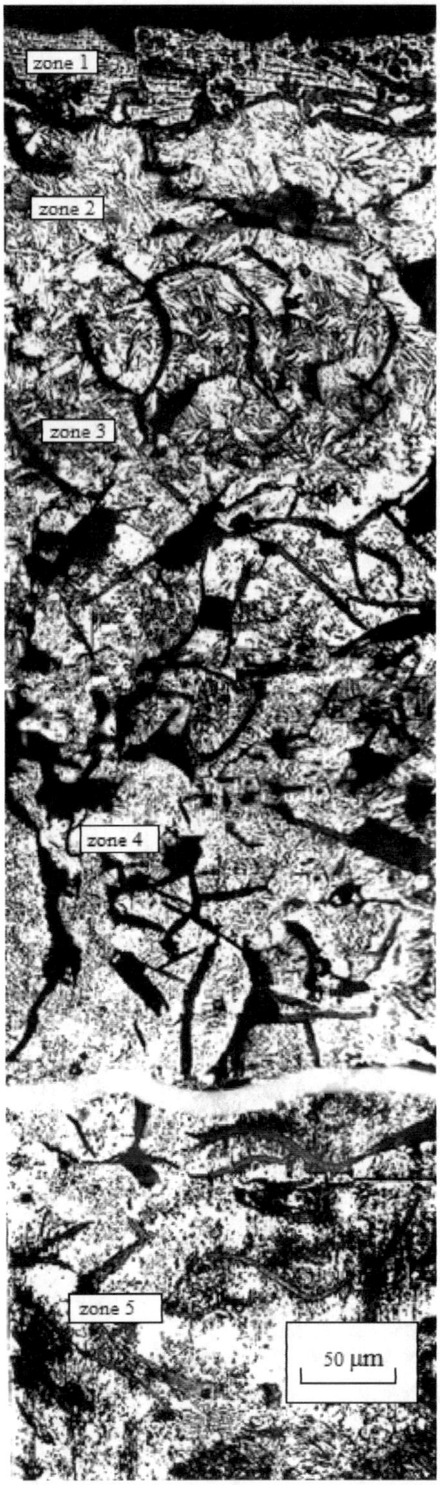

Figure 5. Structure of hardened layer after plasma hardening (q=3300 W/cm^2, k=10 cm^{-2}, V=1.06 cm/s)

The sizes of underlying zones (2-5) at combined processing were increased on the average at 10% and their structure corresponds to structures obtained at plasma heating. The measurement of a surface roughness has shown immutability it at a level created by ultrasonic plastic deformation (Ra 0.66 μm)

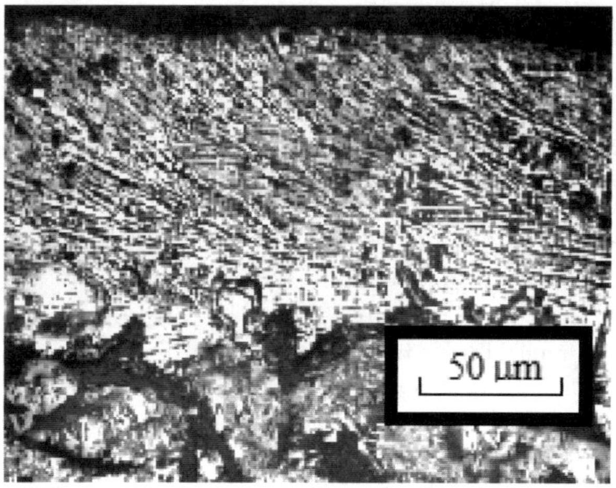

Figure 6. Structure of a hardened layer after combined processing (P=250 N, q= 3300 W/cm^2, k=10cm^{-2}, V=1.06 cm/s)

4. Conclusions

The application of rod approximation for the description of processes of loading at ultrasonic deformation and high-speed heating in combination allowed to execute simulation of process of residual stresses formation. Setting the optimum modes ensures required ratio of residual deformation of a surface with the residual stresses.

The combined hardening-finishing processing of gray cast iron on the base of deformation and thermal effects combination allows to change a kinetics of phase transformation to increase homogeneity of a structure. It makes it possible to get surface layers of an increased thickness with increased hardness without surface melting.

References

[1] Tyurin A, Nagavkin S, Malikov A, Orishich A. 2015 Microstructure of WC-Co hard alloy surface after laser treatment *Surface Engineering* **31 (1)** 74-7

[2] Iskhakova G A, Rakhimyanov Kh M 1987 Investigation of the microstructure and mechanical properties of steel 45 after plasma hardening *Soviet surface engineering and applied electrochemistry* **5** 28-32

[3] Mordyuk B N, Prokopenko G I 2007 Ultrasonic impact peening for the surface properties management *Journal of Sound and Vibration* **308 (3-5)** 855-66

[4] Rusinko A 2011 Analytical description of ultrasonic hardening and softening, *Ultrasonics* **51(6)** 709-14

[5] Klimenov V A, Vlasov V A, Borozna V Y, Klopotov A A 2015 Ultrasonic surface treatment of titanium alloys. The submicrocrystalline state *IOP Conference Series: Materials Science and Engineering* **91 (1)** 012040

[6] Solonenko O P, Alkhymov A P, Marusin V V, Orishich A M, Rakhimyanov Kh M, Salimov R A, Shchukin V G, Kosarev V F et al 2000 *High-Energy Processes of Materials Treatment* 425.